Edgar M. Williams

June, 1992

Joachim Schubert

Dictionary of Effects and Phenomena in Physics

Distribution:
VCH Verlagsgesellschaft, P.O.Box 1260/1280, D-6940 Weinheim
(Federal Republic of Germany)
USA and Canada: VCH Publishers, Suite 909, 220 East 23rd Street, New York NY 10010-4606
(USA)

ISBN 3-527-26676-3 (VCH Verlagsgesellschaft)
ISBN 0-89573-487-7 (VCH Publishers)

Joachim Schubert

Dictionary of Effects and Phenomena in Physics

Descriptions
Applications
Tables

Dipl.-Phys. Dr. Joachim Schubert
Siemens AG
Unternehmensbereich Bauelemente
Balanstr. 73
D-8000 München 80

Editorial Director: W. Greulich
Production Manager: M. Nothacker

Deutsche Bibliothek, Cataloging-in-Publication Data

Schubert, Joachim:
Dictionary of effects and phenomena in physics : descriptions, applications, tabl. / Joachim Schubert. –
Weinheim ; New York : VCH, 1987.
Einheitssacht.: Physikalische Effekte
⟨engl.⟩
ISBN 3-527-26676-3 (Weinheim)
ISBN 0-89573-487-7 (New York)

Composition: Mitterweger Werksatz GmbH, D-6831 Plankstadt
Printing: Diesbach Medien GmbH, D-6940 Weinheim
Bookbinding: Josef Spinner, D-7583 Ottersweier
Printed in the Federal Republic of Germany

Preface

The wide acceptance of the book **Physikalische Effekte** in Germany has encouraged publisher and author to prepare an English edition. The general aim of the book – to present physical effects and phenomena to a broad readership – has not been changed. Some new effects have been added to this edition, as well as some references to current Anglo-American literature.

The dictionary in the book, the tables, the cited literature, and the chronology offer

- an overview of approximately 400 effects and phenomena, their names and their relationships to other effects,
- a list of available and pertinent literature,
- a chronology of the effects, and of major physical events,
- information about the discoverers of the effects.

I want to thank those who have been helpful in preparing this book. In particular, I am grateful to H. Rechenberg for his kind assistance concerning historical problems, and K. Bethge and A. Scharmann for their useful criticism and comments.

Readers' suggestions, especially concerning effects that could be added to this book, are welcome.

Munich, March 1987 J. Schubert

Preface

Contents

Introduction

What are Physical Effects?

At one time "effect" (Latin: **effectus**: outcome, consequence, result) referred to some form of work done, i.e. to some form of power (Coriolis 1832); for example, the current effect is the same as the power of an electric current. The term effective power means efficiency.

During the last hundred years, the term effect has been increasingly used for special physical phenomena. On comparing old with new physics books, one finds a marked change in the choice of words. Instead of the terms phenomenon and principle, which were previously used synonymously, the term effect has become increasingly popular since the beginning of this century. Similarly, "experiment" is being used more often as a synonym for effect.

Experiment has a central position in physics. Any hypothesis or theory is tested for its truth by experiment.

Physical effects are the responses of nature to special experiments. In the simplest case, one physical quantity is measured in dependence on another: for example, the temperature dependence of an ohmic resistance or the heat produced by current flow. One usually tries to derive the simplest possible law relating the measured quantities. Direct proportionality or a quadratic dependence between the quantities is characteristic of simple effects and their theoretical explanation.

In a large number of effects, the cause and the action of the effect may be interchanged. In such cases, one has pairs of inverse effects, such as the Seebeck und Peltier effects, the Wiedemann and Wertheim effects, the Dufour and Ludwig-Soret effects, etc. In these effects, the measured quantities are generally directly proportional to one another. The term inverse effect is also common.

In the literature we also find the terms reciprocal and nonreciprocal effects. The Joule effect, for example, is reciprocal, because the resulting heat is independent of the direction in which the electric current flows. The Cotton-Mouton effect is reciprocal, because the magnetic birefringence is independent of the light-propagation direction. On the other hand, nonreciprocal effects are direction-dependent; for example, the Faraday effect.

Other effects are neither reciprocal nor reversible, for example, the Hall effect. Application of the Hall voltage with the control current flowing does not produce the corresponding magnetic field.

Further, there are analogous effects: many new physical phenomena can be first explained in terms of analogies with known ones. For example, the action of an electric current was initially interpreted by analogy with a flowing liquid. Analogy also allows one to transfer the results of the oscillator equation and the wave equation from the purely mechanical case to optics or electromagnetic oscillations and waves. However, the relevant quantities must then be reinterpreted. An example for analogous effects is provided by the Barkhausen and Portevin-Le Chatelier effects.

The individual subject areas are already mingled in classical physics: mechanical, thermal, electrical, magnetic, and optical quantities influence one another. Novel effects, arising from the quantum properties of matter, become dominant in the atomic field. These either occur directly as macroscopic quantum effects, such as superconductivity and superfluidity (the so-called Onnes effect), or else, are observed in experiments under extreme conditions, such as strong magnetic fields and low temperatures (for example, the Einstein-de Haas effect, the von Klitzing effect or size effects), or are found in the examination of new materials. Other effects are indicative of the quantum nature of particles in the atomic region.

The following two tables (Tables 1 and 2) present a brief overview of the various groups of effects.

In Table 1, the classical effects are arranged in a matrix. The quantities (mechanical, thermal, electrical, magnetic, and optical) are all dependent on one another. Not all positions are filled. Reciprocal or inverse effects appear in the matrix in several places, whereas non-reciprocal effects just appear once.

Table 2 links the effects derived from classical quantities with the new effects related to the quantum features of nature. These new effects can be divided into atomic and quantum effects, scattering effects, etc. Precise explanations of the classical effects are also obtained by way of the quantum properties of matter. At the same time, Table 2 shows some of the major areas of physical research in universities and industry.

In 37 tables (Table 19 to Table 55) covering various fields in physics, the physical effects are arranged in terms of the scheme cause-effect-name. For each field, there is a symbol consisting of up to three bold-face letters. The same symbols are employed in the descriptions of individual effects in the Dictionary of Effects (p. 1 – 97). A list on the inside back cover shows them in alphabetical order and correlates them with the tables. This list enables the reader to establish the arrangement of the effects within the general framework of physics.

Without a basic understanding of physical effects, no progress is possible in applications in engineering and science. Material problems often prevent the use of effects which have been known for some time. An example is the Hall effect,

Table 1. "Classical" effects

Dependent quantities	Mechanical quantities	Thermal quantities	Electrical quantities	Magnetic quantities	Optical quantities
Mechanical quantities	Mechanical effects flow effects acoustic effects	e.g. After effects	Electrokinetic effects electro-mechanical effects	Magnetomechanical effects gyromagnetic effects	Relativistic effects
Thermal quantities	Thermodynamic and kinetic effects		Low-temperature effects, cross effects, electrothermal effects	Magnetothermal effects	Thermooptic effects
Electrical quantities	Electro-kinetic effects, electro-mechanical effects	Electro-thermal effects, cross effects	Electrical and plasma effects, electrolytic effects current conduction effects	Magnetic effects, Size effects	Photoelectric effects
Magnetic quantities	Magnetomechanical effects, gyromatic effects		Electric and magnetic effects, current conduction effects		
Optical quantities	Optoacoustic effects	Thermooptic effects	Electrooptic effects	Magnetooptic effects	Optical effects

Table 2. Relationships between "classical" and "new" effects

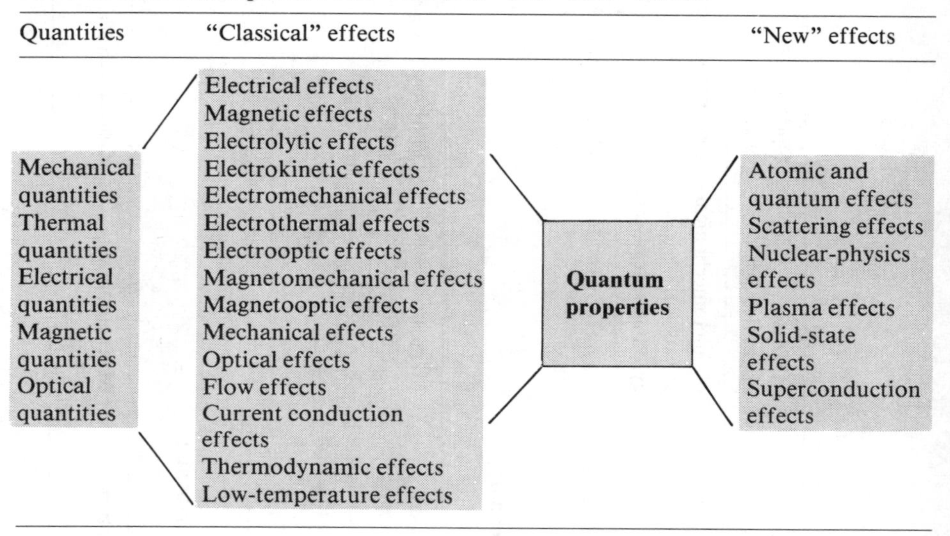

which, today, is being increasingly used in modern semiconductors (so-called III/V compounds).

For this reason, older effects only found in old textbooks have also been included in this book: they await modern applications. Other effects and experiments at present may not possess a technical application, but advance our understanding and influence theory. Examples of this type are provided by the astronomical and relativistic effects, as well as effects in quantum field theory.

The effects described range from those with important applications to those which are mere curiosities.

Although the names given to pyhsical effects are partly accidental, the effects offer a living and informative cross section of physics, since they occur in all physical disciplines.

Notes to the Reader

The effects are described in detail in the *alphabetical Dictionary of Effects* (p. 1 – 97). The significance of the bold-face symbols has already been explained (p. X).

The *tables* (p. 99 – 125) provide guidelines; they are arranged alphabetically and contain the effects occuring in a particular field.

The *literature cited* (e.g. [6]) includes monographs, textbooks, and dictionaries. Special attention was given to standard works written in English which are listed as "additional literature". The original literature could not be traced in all cases, and then the source cited is usually a textbook or a dictionary (Literature: p. 135 ff.).

For the older effects, the historical data are almost complete; this does not always apply to the effects discovered in the postwar years and biographical details of their discoverers were sometimes difficult to locate.

Dictionary of Effects and Phenomena

The literature referred to is given on p. 135 ff.

AC-DC effect

→ Lenard effect (1890)

AC Kerr effect L

If a crystal is pumped with an intense laser beam and at the same time is irradiated by a flash lamp, induced birefringence of the latter light occurs. The effect is an analogue of the classical Kerr effect; the strong steady field is replaced by the very intense pumping beam (AC = alternating current).

→ Kerr effect, → Raman-induced Kerr effect, → nonlinear optical effects

Hellwarth, R.W. *Progr. Quant. Electron.* **5**, 1 (1977)

Lit.: [96, 98, 100]

Acoustoelectric effect SS, Sc

This effect describes the interaction between a sound wave and the quasifree charge carriers in a semiconductor. The wave is usually attenuated as a result of absorption, and energy is transferred to the charge carriers and the lattice. However, it is possible for the sound wave to be amplified in a semiconductor exhibiting piezoelectric properties.

The piezoelectricity (→ piezo effect) in the material produces potential modulation with the advancing sound wave, whereby the charge carriers can accumulate in the potential troughs. If, in addition, a steady electric field is applied, this produces a drift velocity in the charge carriers. When the drift velocity is greater than the speed of sound, the inverse piezoelectric effect produces a correctly phased energy transfer to the sound wave, and thus amplification.

In general, the system is unstable, and acoustic domains are produces.

The effect is used in acoustoelectric amplifiers and in delay lines.

Lit.: [1, 27, 53, 71, 75] (additional lit. see Table 45 and 51)

Adiabatic galvanomagnetic effects

→ galvanomagnetic effects

Adiabatic Hall effect

→ current conduction effects, → cross effects, → galvanomagnetic effects

Aftereffects

→ relaxation effects

Alexandrite effect

The very rare mineral alexandrite is a form of chrysoberyl ($BeAl_2O_4{:}Cr^{3+}$). The chromium ions are responsible for the color in alexandrite, as in ruby. The internal forces are so balanced that the color changes with the illumination (the alexandrite effect). If the blue component predominates, the crystal appears blue-green, while if the light is mainly red, the crystal becomes red.

Nassau, K.: *Gems Made by Man*. Chilton Book Co., Radnor Pa. 1980

Alignment effects G

These effects relate to the alignment of atomic or molecular magnetic moments in the direction of an external magnetic field, and similarly to the alignment of electric dipoles or dipole moments of atoms or molecules in an electric field.

→ Stern-Gerlach effect, → electrooptic effects, → magnetooptic effects.

Lit.: [45] (additional lit. see Table 19)

Alternating-field luminescence

→ Destriau effect

Analogue states N

Analogue states is the name given to energy states in the nucleus that have unequal charge numbers but equal mass numbers in combination with the same isospin but differing third components of the isospin. There is a difference in the energies of such states because of the mass difference between the neutron and proton (*mass effect*) and the difference in the Coulomb energies (*Coulomb effect*). The mass effect and the Coulomb effect have opposite signs. The mass effect is smaller than the Coulomb effect.

Lit.: [46, 75] (additional lit. see Table 38)

Anaphoresis

→ electrophoresis

Anisotropy effect S

The production of Cooper pairs (→ superconductivity) in a superconductor involves phonons (quanta in the elastic lattice vibrations). In an anisotropic crystal (or metal), it is therefore possible for the production of Cooper pairs to be direction-dependent, which is called an anisotropy effect.

Lit.: [11] (additional lit. see Table 44)

Anisotropy effects G

Anisotropy means that there are differences in the physical properties in different directions within a body. Anisotropy is derived from the internal, spatially differing structures in the body. An isotropic body can be made anisotropic by external factors such as pressure, tension, mechanical or electrical strain, magnetic fields, etc. Optical anisotropy results in → birefringence. Anisotropy in a conductor can affect, for example, → galvanomagnetic and thermomagnetic effects.

Lit.: [76], *textbooks, dictionaries (additional lit. see Table 19)*

Anomalous Barkhausen effect

→ Barkhausen effect, anomalous

Anomalous Hall effect

→ galvanomagnetic effects

Anomalous skin effect

→ skin effect, anomalous

Anomalous Zeeman effect

→ Zeeman effect

Arago's experiment (1824) EM

This experiment concerns *eddy-current effects*, which are produced by induction: the magnetic needle that is free to rotate follows the rotation of a copper disc and vice versa. Arago could not explain this and denoted the effect as *rotation magnetism*. The motion, for example, of the magnetic needle induces currents in the disc in accordance with Lenz's law. These are directed in such a way that they oppose the cause producing them. The part of the apparatus originally at rest follows the motion of the other part.

ARAGO, Dominique Francois Jean,
French physicist and politician, 26 February 1786 – 2 October 1853,
Professor in Paris
→ Rowland effect

Lit.: [6b, 50 Vol. 4, part 1) (additional lit. see Table 26)

Archimedes' principle Me

(uplift)

Hydrostatic uplift is the name given to a force opposed to the force of gravity experienced by a body immersed in a fluid at rest. This force is produced by the fluid pressure on the surface.

Archimedes' principle states that the hydrostatic uplift is equal to the weight of the fluid displaced by the body. The force acts through the center of gravity of the displaced volume. The body undergoes an apparent loss in weight equal to the weight of the displaced volume.

This principle can be used to measure density, or, if the density is known, to measure volume.

ARCHIMEDES,
Greek scientist (-287?) – (-212)

Lit.: *textbooks, dictionaries, handbooks*

Asymmetry effect

→ slowing down effect

Auger effect (1926) AQ II

The absorption of X rays by atoms can lead to internal photoionization. The atom then returns to the ground state by radiationless transition of an outer-shell electron. The energy thereby released is used in freeing an electron from the outer shell. The process is also called *internal absorption of X rays or self-ionization*, in which a virtual X-ray quantum is produced, and later reabsorbed on releasing an electron. The presence of the virtual quantum as intermediary is not necessary for the Auger process.

Today, the Auger effect is defined as the fact that an excited atom or molecule or else a defect in a solid gives up its energy not as light but by the emission of an electron. These released electrons are characteristic of the element. They are used in microanalysis of the light elements, e.g., in connection with electron microscopes. The Auger effect also occurs in semiconductors, where it is detectable as interference. In that case, the excitation is produced not by X rays but by injected charge carriers. The Auger effect must be excluded in order to ensure reliable operation of optoelectronic components such as LEDs (light-emitting diodes) or laser diodes.

AUGER, Pierre Victor,
French physicist, 14 May 1899, from 1937 Professor in Paris

Auger, P., *J. Phys. Radium* **6,** 205 (1925); *C. R.* **180,** 65 (1925); *C. R.* **182,** 773 (1926)
Braunbeck, W., *Phys. Bl.* **26,** 264 (1970)

Lit.: [6 d] (additional lit. see Table 21)

Autoelectron emission

→ field effect

Avalanche effect (1953) SCC

If very high potential differences occur across the space-charge zone in a p-n junction, the latter loses its insulating property. The high potential accelerates the charge carriers in the space-charge zone so much that new charge carriers can be produced in an avalanche by collisional ionization. If the semiconductor is highly doped, the → tunnel effect becomes dominant before the avalanche effect can build up. The avalanche effect belongs to the high-field effects observed in semiconductors.

→ Gunn effect, → Zener effect

McKay, K.G., *Phys. Rev.* **91**, 1079 (1953)

Lit.: [63, 74, 75] (additional lit. see Table 49)

Azbel-Kaner effect (1956) SS

(cyclotron resonance effect)

This phenomenon occurs when a conductor bearing a current is placed in a microwave field (GHz range) and a magnetic field is applied perpendicular to it (0.02 – 0.5 T); it is also called → the anomalous skin effect. On account of the skin effect, the microwaves penetrate only about 10^{-5} cm into the conductor. The magnetic field causes the penetration depth to be appreciably smaller than the radius of the cyclotron paths of electrons in the magnetic field. Absorption resonance occurs if the electrons can pass repeatedly through the path involved in the microwave skin effect without scattering: the electrons are then accelerated in the microwave field and draw energy from it. The resonance effect is observed at low temperatures (about 4 K). The Azbel-Kaner effect belongs to the group of effects in which a metal is present in a high-frequency electromagnetic field.

→ skin effect, anomalous skin effect, → Sondheimer effect, → size effects

Azbel, M.J. and Kaner, E.A., *SETF* **32**, 896 (1956)

Lit.: [1 a, 5 d, 33 a] (additional lit. see Table 51)

Babinet's principle

Babinet's principle states that the deflection of waves is identical at complementary assemblies consisting of slits or screens, e.g., a slit-wire, a hole-circular disc structure outside the central pattern. Consequently, the diffraction at dust particles is identical with that at many small holes. The waves produced by complementary obstacles have opposite phases.

Babinet's principle can be used, for example, in the calculation of antennas and in high-frequency engineering, in general.

BABINET, Jacques,
French physicist, 5 March 1794 – 21 October 1872, Professor in Paris

Lit.: [6 c, 9] *dictionaries, handbooks*

Back-Goudsmit effect (1927) Mo

In the United States, the → Paschen-Back effect of the hyperfine structure is called the Back-Goudsmit effect: a strong magnetic field uncouples the nuclear and shell-electron spins, just as the Paschen-Back effect involves uncoupling between the orbital angular momentum and the spin. The energy of the magnetic interaction between the shell and the magnetic field must be greater than the energy of the hyperfine interaction between the shell and the nucleus.

The effect is frequently called the Paschen-Back effect of hyperfine structure, whereas Back-Goudsmit effect is sometimes used as the American name for the Paschen-Back effect.

BACK, Ernst
German physicist, 21 October 1881 – 20 July 1959

GOUDSMIT, Samuel Abraham,
Dutch-American physicist, 11 July 1902 – 4 December 1978, in the USA since 1927, finally Professor in Upton (USA)

Goudsmit, S. and Back, E., *Z. Phys.* **43**, 321 (1927)

Lit.: [7, 46, 47] (additional lit. see Table 37)

Ballo electricity

→ Lenard effect

Barkhausen effect (1919) EM

(Barkhausen noise, Barkhausen steps)

Barkhausen found that on magnetization reversal in iron, the switching in the elementary magnetic domains produced a noise signal in a coil. This can be detected if an induction coil with very many turns is mounted on an iron core. An amplifier enables one to hear the elementary switching processes. When the external magnetic field is altered, the energetically most favored spontaneous-magnetization areas (Weiss domains) enlarge by rotation and wall shifts. Depending on the local situation in a domain, changes in field strength lead to reversible wall shifts or to sudden irreversible movements: the Barkhausen steps. When the field changes rapidly, the superposition of many steps gives rise to Barkhausen noise, while when the field changes slowly, the individual magnetization reversal pulses can be detected by ear or eye (with an oscilloscope). The Barkhausen effect is the magnetic analogue of the → Portevin-le Chatelier effect.

BARKHAUSEN, Heinrich Georg,
German physicist, 2 December 1881 – 20 February 1956, from 1911 Professor in Dresden
Barkhausen, H., *Phys. Z.* **20**, 401 (1919)
Lit.: [37, 74, 75, 76] (additional lit. see Table 26)

Barkhausen effect, anomalous EM

Anomalous Barkhausen effect is the name given to the occurrence of large steps in the magnetization of an iron-aluminum alloy above 400° C. The effect is ascribed to the orientation of Al-Al pairs, which stabilize the domain walls. The effect is temperature- and frequency-dependent.
Lit.: [37] (additional lit. see Table 26)

Barkhausen effect in thin films EM

The frequency of Barkhausen steps decreases exponentially with the thickness of thin films. In a thin film magnetization reversal involves nucleation difficulties if the film edge is shadowed by a mask during evaporation and therefore is thinner than the middle. This is called the *slope effect*. The thin film edge makes nucleation more difficult. When the field is sufficiently strong, nuclei are formed in the center, and the film is finally reversed in magnetization in a single large Barkhausen step.
Brackmann, J., Küsterer, W., and Netzdorf, W., *Z. Angew. Phys.* **13**, 313 (1961)
Lit.: [37] (additional lit. see Table 26)

Barkhausen noise

→ Barkhausen effect

Barkhausen step

→ Barkhausen effect

Barnett effect (1915) EM

(Richardson-Barnett effect)
The Barnett effect describes the relationship between the magnetization and the rotation of a magnetized rod. In his experiment Barnett used an iron rod, which he rotated very rapidly about its longitudinal axis. In addition to its spin, each free electron then has a mechanical angular momentum, which is associated with a magnetic moment acting in the direction of the rotation axis of the body. The body therefore exhibits magnetization. The Barnett effect is observable only in ferromagnetic materials.
→ Einstein-de Haas effect

BARNETT, Samuel Jackson,
American physicist, 14 December 1873 USA – ?
Barnett, S.J., *Phys. Rev.* **6**, 239 (1915)
Barnett, S.J., *Rev. Mod. Phys.* **7**, 129 (1935)
Lit.: [50 Vol. 4, part 4, 74, 75, 76] (additional lit. see Table 26)

Barotropic phenomenon (1906) TK

In a binary mixture in which the two substances have very different critical temperatures, it can happen that when the liquid and vapor phases coexist, the vapor phase has the greater density. Then the gas (the vapor) sinks below the liquid surface (barotropic phenomenon). The phenomenon was discovered in 1906 by Kamerlingh-Onnes in a mixture of He and H (1:6) at 49 bar at the boiling point of hydrogen.

KAMERLINGH-ONNES, Heike,
Dutch physicist, 21 September 1853 – 21 February 1926
Lit.: [76, 77] (additional lit. see Table 54)

Barrier or barrier-layer effect

→ rectifier effect

Barrier-layer photoelectric effect PS

(Adams and Day 1876, Grondahl 1926, Lange and Schottky 1930)
(photovoltaic effect, p-n photoelectric effect)
In principle, this concerns effects occurring in photoconductors on illumination. The electrons and holes in the barrier layer of a p-n junction are separated by an internal field. A distinction is drawn between the *transverse* and *longitudinal p-n photoelectric effects,* in accordance with the direction in which the light is incident on the semiconductor. In transverse incidence on a p-n

junction, the voltage-current line of the diode is displaced as a whole, with the photocurrent linearly dependent on the intensity. Also, the photodiode is light-sensitive only in a narrow region on transverse illumination. This disadvantage is avoided in longitudinal illumination. The irradiated barrier layer acts as a current and voltage generator; a photovoltaic potential difference appears at the open terminals without an external voltage: *photovoltaic operation.* On short circuit, a short-circuit current is obtained. If a bias voltage is applied to the p-n junction, one speaks of *photoelectric operation.* The diode operates as a current generator with a high internal impedance.
The effect occurs only in the region of the fundamental lattice absorption.
Adams and Day, *Proc. R. Soc. London* **25,** 113 (1876)
Braun, F., *Wied. Ann.* **4,** 476 (1878)
Grondahl, L. O., *U.S. Pat. 1640335* (1925)
Geiger, Ph., *Brit. Pat. 277610* (1927)
Lit.: [2, 6d, 28, 50 Vol. 4, part 4, 91, 92] (additional lit. see Table 42)

Bauschinger effect (1886) So

If a material is plastically deformed first in one direction and then in the opposite direction, the elastic limit may be reduced under certain circumstances. In the ideal Bauschinger effect, the yield stress in compression is reduced by the amount by which the yield stress in tension increases. This is due to the formation of elastic stresses opposite in direction to the original shape deformation.
The effect plays a part in materials testing, e.g. in measuring the elastic properties of steel. Repeated torsion applied to a body in one direction and then in the other enables one to evaluate the mechanical properties of the sample from the *Bauschinger torsion diagram.*

BAUSCHINGER, Johann,
German engineer, 11 June 1834 – 25 November 1893
Heyn, E., *Naturwissenschaften* **9,** 331 (1921)
Furr, S., *Mater. Eng.* **6,** 77 (1973)
Lit.: [74, 75, 76] (additional lit. see Table 50)

Beck arc

→ Beck effect

Beck effect (1911) EM

(Beck arc)
If the anode carbon in a high-current arc is treated with rare-earth salts, e.g. cerium salts, there is a considerable increase in the light intensity when a certain current is exceeded. A pronounced kink in the current-voltage characteristic is observed. The potential increase of a few volts is thought to be due to the evaporation of the salts.

BECK, Heinrich,
German electrical engineer, 20 September 1778 – 17 August 1937
Finkelnburg, W.: *High-Current Carbon Arcs* [in German]. Springer, Berlin 1948
Lit.: [72, 73, 74, 75, 76, 85, 87 Vol. XXII] (additional lit. see Table 26)

Becquerel effect (1839) P

In 1839, E. Becquerel discovered that if one of two similar electrodes immersed in an electrolyte was illuminated, under certain circumstances a potential difference arose between the electrodes. The mechanism has not been established with certainty.
Also, it is by no means sure that the effect is based on photoelectric action, and not on some photochemical action without electron emission. The effect cannot be explained in terms of heating. The potentials are dependent in a complicated and often irreproducible fashion on the light intensity and wavelength. The potentials arise within about a minute and attain a value of about 10^{-4} V. Even the sign of the voltage is not determined for a given arrangement, since under certain circumstances it can reverse during illumination. The effect has occasionally been used in radiation measurements, but not with much success. The Becquerel effect can be taken as the earliest observation of a photo effect.

BECQUEREL, Alexandre Edmond,
French physicist, 24 March 1820 – 11 May 1891, Paris
Becquerel, A.E., *Ann. Chim. Phys.* **71,** (1839)
Becquerel, A.E., *C.R.* **9,** 561 (1839)
Becquerel, A.E., *Ann. Chim. Phys.* **9,** 257 (1843)
Lit.: [50 Vol. 4, part 4] [72, 74, 75, 76, 86] (additional lit. see Table 40)

Becquerel effect (1906) Mo

(paramagnetic rotation)
J. Becquerel examined the → Faraday effect in paramagnetic substances in 1906 (paramagnetic rotation).

BECQUEREL, Jean Antoine Edmond Marie,
French physicist, 5 February 1878 – 4 July 1953
Becquerel, J., *Phys. Z.* **8,** 929 (1907)
Becquerel, J., *Phys. Z.* **9,** 94 (1908)
Becquerel, J., *C.R.* **144,** 682 (1907)
Becquerel, J., *C.R.* **143,** 769 (1906)
Becquerel, J., *C.R.* **145,** 765 (1907)
Becquerel, J., *C.R.* **148,** 158 (1909)
Lit.: [3, 27a Vol. II, 2.2] (additional lit. see Table 37)

Bénard effect Sy

When a planar layer of fluid is heated from below, a temperature difference arises between the bottom surface and the top one. If this difference is small, the layer of liquid remains at rest and the heat is transported by conduction. When the temperature difference exceeds a critical value, hexagonal cells arise (Bénard instability, Bénard effect, Bénard convection), and polygonal rolls occur at larger temperature differences.
The Bénard effect resembles a laser since they both provide classic examples of collective or → synergetic phenomena, in which, far from equilibrium, highly ordered structures can arise out of chaos.
→ synergetic phenomena
Haken, H.: *Synergetics*. Springer, Berlin-Heidelberg-New York 1983
Lit.: [102a, 103b, 112a, 116a]

Benedicks effects, 1st and 2nd

→ thermoelectric effects

Berremann effect (1963) PS

If light falls on a plate that is substantially thinner than the wavelength of the light, only the excitation of the lowest induced polaritons is observed. *Polaritons* resemble *excitons* in being quasiparticles in solids that are formed as the result of strong interaction between photons and nonlocalized excited states. Excitons occur in the production of electron-hole pairs in semi-conductors or solids. The electron and hole move together and can transport energy but not charge. A photon and an exciton then represent a new system if the interaction energy corresponds to the photon energy: this polariton behaves as a new particle, whose dispersion curve is composed of two branches which do not correspond to the dispersion curves for the individual particles.
If the frequency of the induced polaritons coincides with the frequencies of the longitudinal and transverse bulk phonons, one speaks of the Berremann effect.
Lit.: [78] (additional lit. see Table 42)

Bethe effect

→ Lamb shift

Bhabha scattering

→ particle scattering

Binaural effect Me

The ear is capable of determining the direction of a sound source from the difference in the times of arrival of the sound at the two ears. The resolution is about 30 μs. The sound source appears to lie to one side of the median plane. With a delay of about 630 μs, the sound is perceived as coming from the side (90°).
The effect is of importance in the stereo reproduction of radio transmissions.
Lit.: [50 Vol. I, 3] [76, 85]

Bioluminescence

→ luminescence effects

Birefringence G

Birefringence is the name given to the phenomenon in which a light wave incident on an optically anisotropic medium is split into two partial waves with different propagation directions. The partial waves have different normal directions and different phase velocities (normal velocities). The two partial waves in a nonabsorbing medium are linearly polarized at right angles to one another.
A birefringent crystal is of noncubic symmetry. A distinction is also made between optically uniaxial and optically biaxial crystals. In an optically uniaxial crystal, the speed of one partial wave, the *ordinary ray*, is independent of direction. The other partial wave is called the *extraordinary ray*. In the case of the extraordinary

ray, the ray velocity does not agree with the normal velocity. Snell's law of refraction applies with the appropriate refractive index for each of the rays.
If the ordinary ray propagates faster (more slowly) than the extraordinary one, the crystal is called uniaxial positive (negative). The birefringence in uniaxial crystals is used to produce linearly polarized radiation.
In addition to natural birefringence in optically anisotropic crystals, artificial birefringence can be produced by external means in normally isotropic media: *compression birefringence, extension birefringence, flow birefringence, electrical birefringence* (→ electrooptic effects), and *magnetic birefringence* (→ magnetooptic effects).
Lit.: [6c, 22c, 50 Vol. II, 1, 76] (additional lit. see Table 19)

Bistability effect
→ liquid-crystal effects

Blears effect
When organic vapor is present in the vacuum, the signal from an ionization-gauge tube is dependent on the geometry of the system with respect to the working volume to be measured. The effect can falsify the result from the measurement up to an order of magnitude.
Blears, J., *Nature* **154**, 20 (1944)
Blears, J., *Proc. R. Soc. London, Ser. A* **188**, 62 (1946)

Bobeck effect (1967) EM
(magnetic bubbles, bubble memory)
In the absence of a magnetic field, magnetic strip domains may be observed in a thin magnetic film with the easy axis oriented perpendicular to the plane of the film. In 1967, Bobeck showed that these domains contract to cylindrical domains, so-called *magnetic bubbles*, when an external magnetic field is applied perpendicular to the film.
In a magnetic field, the strip domains unfavorably magnetized with respect to the field shrink in favor of the better-oriented ones. At a certain field strength, these become cylindrical domains (domains or bubbles), which vanish by collapse at higher field strengths.
The magnetic cylinder domains or magnetic bubbles in a defect-free single-crystal film (preferably made from a material with a garnet structure) are so mobile that they will move in the leakage field of nickel-iron film components in a magnetic field rotating in the film plane. Consequently, these single-crystal films can be used as magnetic memory material for nonvolatile data. Bubble memories have subsequently (1979) reached a capacity of 256,000 bit or 1 million bit per chip.
Tests are currently being made to see if the → Faraday effect can be used to apply the bubbles in displays.
Bobeck, A.H. et al., *IEEE Trans. Magn.* **5**, 544 (1969)
Lill, A., *Phys. Unserer Zeit* **5**, 69 (1974)
Lit.: [18, 26] (additional lit. see Table 26)

Boersch effect
On emission from the cathode the electrons in a Wehnelt cylinder have differing velocities on account of the Maxwellian velocity distribution. This distribution is broadened by the production of a space-charge region in front of the cathode. The deviation of the electron energy distribution from the Maxwellian is called the Boersch effect, and its size is of the order of an electron-volt. The Boersch effect is of significance in electron microscopy and electron-beam lithography, since it affects the energy and focussing of electrons, and therefore the resolution of the microscope or of electron-beam lithography.
BOERSCH, Hans Paul,
German physicist, 1 June 1909
Lit.: [62]

Bohm-Aharonov effect EM
In electron diffraction, the electron phases may be affected by a magnetic flux. This effect can be detected by the change in the interference pattern. According to Bohm and Aharonov, the phase shift is produced by the vector potential related to the flux. The effect can be used in the measurement of very small magnetic fluxes, e.g., in superconducting hollow cylinders.
The extent of the trapped magnetic flux may be estimated from the interference pattern.
BOHM, David Joseph,
American physicist, 20 December 1917
AHARONOV, Yakir,
Israeli-American physicist
Lit.: [6c] (additional lit. see Table 26)

Böning effect EM

The Böning effect is the displacement of associated ions bound to capturing ions in fine channels in a dielectric medium in response to an electric field. The displacement affects the dielectric properties of the medium.

BÖNING, Paul Johannes,
German physicist and electrical engineer, 6 October 1887 – ?

Lit.: [85] (additional lit. see Table 26)

Born scattering

→ particle scattering

Borrmann effect SS

In X-ray photography, which is an X-ray diffraction method of examining single crystals, Borrmann effect is the name given to the anomalous (very slight) absorption of X-rays in a single crystal. The effect cannot be explained in terms of the mass attenuation law. This anomalously low absorption can only occur when there is at least one diffracted beam. The dynamic theory of X-ray diffraction by crystals describes the effect quantitatively. An analogous effect is observed in electron and neutron diffraction. The method allows one to examine, for example, silicon wafers of a thickness of several millimeters.

BORRMANN, Gerhard Heinrich,
German physicist, 30 April 1908

Eutin, I.T., *Phys. Status Solidi B* **90,** 2, 575 (1978)

Lit.: [25, 75] (additional lit. see Table 51)

Bragg effect L

In the recording of white-light holograms (→ holography), the photographic film of thickness 5 – 15 μm is irradiated from both sides: by the object beam from one side and the reference beam from the other. Standing waves arise in the photographic material and after development produce blackening in film planes separated by $\lambda_0/2$.

In white-light reconstruction, reflections occur at these planes in the same way as reflection of X-rays at the planes in a body.

The Bragg effect in holography is identical with the reflection of X-rays from grating planes; in holography it is sometimes also called the "color filter" effect.

→ holography
→ Laue effect

Lit.: [33] (additional lit. see Table 31)

Branley-Lenard effect P

Lenard confirmed an effect discovered by Branley, namely, that ultraviolet light (120 – 150 nm) produces strong ionization in air and other gases. The gas conductivity is increased. The resulting ion pairs differ in mobility. It is assumed that the photoelectric effect produces ionization in the gas via heavy dust particles present.

BRANLEY, Edouard,
French physicist, 23 October 1844 – 24 March 1940

Lit.: [73, 76] (additional lit. see Table 40)

Breaking effects (1923) El

In a strong electrolyte, the speed with which the ions drift is controlled by two effects: the electrophoretic effect (Debye and Hückel) and the relaxation effect.

Each ion moves in opposition to an oppositely charged ion cloud. In addition, the ion cloud contains water of hydration, so the ion moves against a liquid current. The drift velocity appears to be reduced by the amount by which the ion cloud moves. This drift inhibition is called the *electrophoretic effect*, or the longitudinal or cataphoretic effect (→ electrokinetic effects). The viscosity of the solvent also plays a part.

When a voltage is applied, the finite relaxation time for the ion clouds results in an unsymmetrical distribution of the ion cloud around an ion. This is the *asymmetry* or → *relaxation effect*. The asymmetry leads to a reduction in the ion mobility.

The two braking effects are the more marked the lower the temperature and the higher the concentration.

Debye, P. and Hückel, E., *Phys. Z.* **24,** 305 (1923)

Onsager, L., *Phys. Z.* **27,** 388 (1926); **28,** 277 (1927)

Debye, P. and Falkenhagen, H., *Phys. Z.* **29,** 111, 401 (1928)

Lit.: [24, 35, 50 Vol. 4, part 4] [86 Vol. XIII, 90, 94, 95, 96] (additional lit. see Table 25)

Bridgman effect

→ thermoelectric effects

Bridgman heat

→ thermoelectric effects

Brillouin scattering (1922) Sc

(Brillouin effect)

In 1922, Brillouin examined the scattering of light at statistical fluctuations in the densities of solids, gases and liquids. In addition to the Rayleigh-scattered light, he found a doublet whose components were shifted upwards and downwards by a characteristic frequency: the frequency of the Brillouin light is raised or lowered by the frequency of the sound with which the fluctuations propagate in the medium. The incident photon either produces a phonon (*Stokes scattering*) or annihilates one (*anti-Stokes scattering*) in the solid. In the first case, energy is given up, but in the second, energy is taken up. If this is an acoustic-branch phonon (acoustic phonon), one speaks of Brillouin scattering, while for an optical-branch phonon (optical phonon) one speaks of Raman scattering.

Basically, this is the → optical Doppler effect produced by a sound wave. The observed frequency shift is minimal and much smaller than that in Raman scattering. The → Debye-Sears effect is a special case of Brillouin scattering.

If high light intensities are used, induced Brillouin scattering is produced.

→ Raman effect, → Debye-Sears effect,
→ scattering effects

BRILLOUIN, Léon Nicolas,
French physicist, 7 August 1889 – October 1969
Lit.: [38 Vol. IX] (additional lit. see Table 46)

Bulk effect G

This is a effect occurring in the bulk of a solid that is derived directly from the properties of the material. In contrast, there are surface effects, such as the photoelectric or light-electric effect, or effects at junctions.

For example, the → Gunn effect is a bulk effect.
→ isotope effect
Lit.: [74] (additional lit. see Table 19)

Burstein effect (1954) SC

At high carrier densities in a semiconductor, the absorption edge shifts to higher energies because the states involved in the absorption near the band edges are already occupied. High carrier densities can be produced by injecting charge carriers or by high-level doping with donors or acceptors. The effect is of significance for light-emitting diodes and for laser diodes.

BURSTEIN, Elias,
American physicist, 30 September 1917
Burstein, E., *Phys. Rev.* **93,** 632 (1954)
Lit.: [27, 48] (additional lit. see Table 48)

Bypass excitation

→ Renninger effect

Cage effect El

If the products from the dissociation of molecules cannot diffuse apart rapidly because of other molecules, the reverse reaction may occur. This is called the cage effect.
Lit.: [78] (additional lit. see Table 25)

Casimir effect (1948, 1956) AQ II

Bonding between neutral atoms is brought about by (attractive) van de Waals forces, Eisenschitz and London (1930) calculated a potential proportional to r^{-6} for these forces from quantum mechanics (r is the separation between the atoms). When r is of the order of or greater than λ, where λ is the wavelength for the transition from the ground state to the first excited state, the potential becomes proportional to r^{-7} (Casimir and Polder, 1948). Casimir later gave a simplified derivation which took the zero-point energy into account. He and others also calculated the corresponding attractive forces between two ideally conducting plates and between a plate and a cylinder. The effect between two plates, which is proportional to d^{-4} (d is the plate separation), was first measured reliably in 1956 (Deriagin, Abrikosova, and Lifshitz; Kitchener and Prosser).

CASIMIR, H.B.G.,
Dutch physicist, 15 July 1909, 1933 – 1942 Professor in Leyden, from 1942 at Philips in Eindhoven
Casimir, H.B.G. and Polder, D., *Phys. Rev.* **73,** 360 (1948)
Liftshitz, E.M., *Soviet Physics JETP* **2,** 73 (1956)
Deriagin, B.V., Abrikosova, I.I., and Lifshitz, E.M., *Quarterly Rev.* **10,** 295 (1956)
Prosser, A.P. and Kitchener, J.A., *Nature* **178,** 1339 (1956)
Lit.: [95, 96, 100, 111, 114, 121]

Cataphoresis effect

→ electrophoresis, → electrokinetic effects

Cataphoretic effect

→ braking effects

Cathodoluminescence

→ luminescence effects

Chanelling effect SS

In a crystal lattice, incident ions such as protons move predominantly between the ion-bearing planes, as in a channel. Proton scattering at one or more lattice atoms provides information on the crystal structure: a proton can be scattered into a channel and pass freely through it, or else scattering can occur in the direction of densely occupied lattice planes. Then the proton can no longer pass through the crystal. The effect is used in proton-scattering microscopy in the structural analysis of crystals.

Lit.: [75] (additional lit. see Table 51)

Chemiluminescence

→ luminescence effects

Cherenkov effect (1934) AQ II

When a charged particle moves at high speed through a dielectric medium, a characteristic radiation is produced, mainly in the blue part of the spectrum. The effect occurs when the speed of the particle in the dielectric medium is greater than the phase velocity of light in the medium. In that case, a conical radiation front is produced, similar to the Mach cone in the propagation of sound. It is found that the vertex angle bears a simple relation to the refractive index.

The Cherenkov effect can be used in detecting and counting high-energy particles and also in measuring their speed.

CHERENKOV, Pavel Aleksevich,
Russian physicist, 28 July 1904,
from 1936 Professor in Moscow

Cherenkov, P.A., *Dokl. Akad. Nauk. SSSR* **2,** 547 (1934)
Tamm, I.E. and Frank, I.M., *Dokl. Akad. Nauk. SSSR* **14,** 107 (1937)
Čerenkov, P.A., *Phys. Rev.* **52,** 379 (1937)

Lit.: [22d, 17, 66, 72 – 76] (additional lit. see Table 21)

Circular magnetostriction effect

→ Wiedemann effect

Coanda effect F

The original direction of a freely flowing fluid is deflected by a solid surface nearby. The symmetry of the secondary air flow associated with the liquid flow is perturbed by the surface, which leads to an adaptation of the flow to the wall. The Coanda effect provides the basis for fluidics (hydrodynamic control circuits).

COANDA, Henri Marie,
Rumanian aeronautical engineer, 6 June 1885 – 1972

Lit.: [72]

Coincidence effect

→ track adaption effect

Cold-conductor effect SCC

By cold conductor in the narrow sense, one means a material that shows a sudden increase in resistivity by up to seven orders of magnitude as the temperature increases over a narrow range (positive temperature coefficient, PTC thermistor).

The effect is particularly pronounced in semiconducting ferroelectric titanate ceramics. The temperature of the steep rise can be adjusted, for example, to between 0 and 350° C, by varying the composition.

The resistance anomaly is due to the production of potential thresholds at the crystal boundaries when the ferroelectric Curie temperature of the material is exceeded.

The cold-conductor effect has been used in exact temperature measurement, for control purposes, and in the production of self-thermostatic heat sources.

Kniepkamp, H. and Heywang, W., *Z. Angew. Phys.* **6,** 385 (1954)

Lit.: [58, 84, 85, 91, 92] (additional lit. see Table 49)

Colligative effects TK

(osmotic effects)

The chemical potential and osmotic pressure are closely related. As the chemical potential is also dependent on other quantities, the changes produced by the osmotic pressure are called osmotic effects, or more recently colligative effects. Examples of such effects are the *molar boiling-point elevation* and the *molar freezing-point depression*.

The *osmotic pressure* is the equilibrium pressure in a two-component (or multicomponent) heterogeneous system which is composed of two phases separated by a diaphragm (or at least a semipermeable membrane).
Lit.: [76] (additional lit. see Table 54)

Color filter effect

→ Bragg effect

Comovement effect AQ I

In the quantum-mechanical calculation of atomic energy terms, it is usual to take the first approximation as the nucleus at rest. In fact, however, the nucleus and the electrons move about their common center of gravity. Consequently, the electron binding energies in the shells must be corrected for this comovement of the atomic nucleus. While the effect is simple to calculate for hydrogen, severe calculation problems arise when there are several shell electrons.
Lit.: [75] (additional lit. see Table 20)

Compensation effect

→ Jaccarino-Peter effect

Compton effect AQ II

In the scattering of short-wave X-rays at electrons in atoms, the original radiation is accompanied by a deflected longer-wavelength radiation dependent in a characteristic fashion on the scattering angle. The relationship between the scattering angle and the wavelength change can be derived directly from the relativistic relation between energy and momentum for particles.
In 1925 Bothe and Geiger demonstrated the validity of the energy and momentum law by means of the Compton effect in individual scattering acts. The Compton effect is a consequence of the particle properties of light (gamma rays).

COMPTON, Arthur Holly,
American physicist, 10 September 1892 – 15 March 1962, Professor in St Louis and Chicago
Compton, A.H., *Phys. Rev.* **21,** 483 (1923)
Debye, P., *Phys. Z.* **24,** 161 (1923)
Lit.: [6d 7, 16, 17, 22d, 27, 38 Vol. III, Vol. IV, 49, 50 Vol. II 2 part 2, 70 Vol. II, 72 – 79, 81, 85, 90] (additional lit. see Table 21)

Compton effect, inverse AQ II

The scattering of γ quanta at fast electrons is called the inverse Compton effect. The electron loses energy, while the frequency of the γ quantum is shifted upwards.
→ Nikischow effect, → scattering effects
Lit.: [6c] [14] (additional lit. see Table 21)

Cooling effect

(in electron emission)
Energy removed from a heated body in electron emission and leading to cooling consists of two parts: the electron work function and the kinetic energy that the electrons have after leaving the body.
Accurate measurements made for tungsten demonstrate that there is good agreement between the work function calculated from the cooling effect and that calculated with the help of Richardson's law.
→ Richardson effect, → shot effect

WEHNELT, Artur Rudolph Berthold,
German physicist, 4 April 1871 – 15 February 1944
Wehnelt, A. and Jentzsch, F., *Verh. Dtsch. Phys. Ges.* **10,** 605 (1980); *Ann. Phys.* **28,** 537 (1909)
Lit.: [50 Vol. 4, part 4] [86]

Cooper effect

→ superconductivity

Corbino effect

→ galvanomagnetic effects

Coriolis force (1832) Me

An additional force acts on a body moving in a rotating frame of reference. It operates perpendicular to the plane passing through the direction of the velocity and the rotation axis. The action of the Coriolis force is such that a moving body in the Northern hemisphere is deflected to the right, whereas in the Southern hemisphere it is deflected to the left. The Coriolis force is therefore of significance in ballistics, canal design, etc.
→ Eötvös effect

CORIOLIS, Gaspard Gustave de,
French physicist and engineer, 21 May 1792 – 19 September 1843, Professor in Paris
Coriolis, G.G. de, *Mém. Sav. étrang* III (1832)
Lit.: [6a, 22a], *dictionaries, textbooks*

Corona effect EM

(St Elmo's fire, Elia's fire)
The arborescent electrical discharges from the tops of highly elevated objects such as lighting conductors and masts are examples of the corona effect. The effect also occurs with electronic components such as capacitors, and can reduce electrical strength. It also leads to additional losses in high-voltage lines.
The effect derives from the strong electric fields arising at sharp curvatures which favor electrons leaving sharp points.
Lit.: [50 Vol. 4, part 1 and 4] [76] (additional lit. see Table 26)

Cotton-Mouton effect (1907) Mo

The magnetic birefringence in a liquid containing anisotropic molecules is called the Cotton-Mouton effect. The effect is also called *magnetic birefringence*. It is the magnetic analogue of the electrooptic → Kerr effect. A magnetic field perpendicular to the direction of light propagation in a transparent material produces artificial birefringence. The path difference between the two rays is proportional to the thickness of the layer traversed and to the square of the magnetic field strength. The effect is based primarily on the alignment of the magnetically anisotropic molecules in the magnetic field.
In general, the coefficient of proportionality is extremely small. The effect has been used in the microwave range.

COTTON, Aimé,
French physicist, 9 October 1869 – 16 April 1951, Professor in Toulouse and Paris
Cotton, A. and Mouton, H., *C.R.* **145,** 229, 230 (1907)
Lit.: [9, 26, 50 Vol. 4, part 4] [72 – 76, 89] (additional lit. see Table 37)

Cottrell effect SS

In research on the temperature dependence of the elastic limit for homogeneous mixed crystals, Cottrell found a jerky deformation, which he ascribed to clouds of impurities at crystal defects. A region enriched in impurities is called a Cottrell atmosphere. The interaction between the defects and the atmospheres, and the production of vacancies and interstitial atoms that can form new atmospheres, explain the sudden deformation. The blocking of dislocation lines by foreign atoms is called the Cottrell effect.

COTTRELL, Alain Howard,
British metallurgist, 17 July 1919
Cottrell, A.H.: *Theory of Crystal Dislocations*. Blackie & Son Ltd., London 1964
Lit.: [43, 72, 87 Vol. I and II] (additional lit. see Table 51)

Coulomb effect

→ analogue states

Coulomb scattering

→ particle scattering

Coupling effect

→ proximity effect

Coupling effects AQ I

Isotopic composition plays a role in the atomic spectra of multielectron atoms. The name coupling effect is given to this mass-dependent contribution to the electron binding energy which is added to the → common-movement effect.
In accordance with the electron movement, the nucleus must move, e.g., in the opposite direction, so that the atom itself remains at rest. Hence, the common-movement effect increases or decreases the binding energy.
Lit.: [38 Vol. III, 72] (additional lit. see Table 20)

Coupling effect, quantum-mechanical AQ I

The atoms in a light-emitting gas may interact so strongly that level shifts occur, with the result that a pressure-dependent broadening of the emission line is observed. This is frequently called a quantum-mechanical coupling effect.
Lit.: [38 Vol. III, 72] (additional lit. see Table 20)

Cross effects G

A feature of transport effects is that generalized fluxes, such as of heat, electric current, or in diffusion, are related to generalized forces; the fluxes are then generally linear functions of the forces. Onsager showed that within the scope of irreversible thermodynamics the related coefficients are symmetrical, even in the case of a magnetic field.

Phenomena in which two (or several) transport effects are coupled (including the converse effects) are known as cross effects:
thermal and electrical conductivity: thermoelectric effects;
thermal conductivity and diffusion: the Ludwig, Soret, and Dufour effect;
electrical conductivity and diffusion: electro-diffusion and the converse effect;
electrical and/or heat flow in a magnetic field: the generalized Hall effects and current flow effects;
(→ current flow effects, → electrokinetic effects)
Lit.: [20, 23] (additional lit. see Table 19)

Crystal effects SS

The name crystal effect is given to the dependence of macroscopic collisional and/or absorption cross sections for particles (neutrons) on the crystal structure of the material. The effect plays a part in reactor calculations, since the effective cross sections for neutrons in the relevant materials must be taken into account.
Lit.: [75] (additional lit. see Table 51)

Crystal photoeffect PS

On illumination natural cuprite crystals show a photo-EMF, whose direction is dependent on that of the incident light beam. The resulting current is proportional to the illumination intensity, whereas the voltage tends to a limit. The effect is also observed with transparent zinc sulfide. One assumes that the absorption occurring along the light path produces an electron-concentration gradient, which drives the electrons in the direction of the light.
Lit.: [76] (additional lit. see Table 42)

Current conduction effects CC

(in metals and semiconductors)
Current conduction in metals gives rise to a range of effects of varying strength, depending on the experimental arrangement. These can be divided into two groups. The first group consists of effects occurring with a current flowing in an inhomogeneous conductor or in different homogeneous conductors connected together. In this case, temperature changes are associated with the electrical potentials or voltages. The group name *thermoelectricity* or *thermoelectric effects* is usually employed for these effects (Table 15).
The second group concerns changes in temperature and potential difference in a conductor bearing a current or heat flux in the presence of a longitudinal or transverse magnetic field. These effects are known under the joint names of *galvanomagnetic* and *thermomagnetic effects* or *generalized Hall effects* (Table 16).
The longitudinal effects occurring in a longitudinal magnetic field have not been named, apart from the magnetoresistance effect. The experimental arrangements and the resulting effects are given in the table.
In general, the changes in heat flux, voltage, and temperature are directly proportional to the relevant quantity. They can all be explained in an elementary fashion in terms of the electron motion in the conductor under the influence of temperature differences or heat fluxes, or electric and magnetic fields, which give rise to the Lorentz forces on the electrons. If the directions of the current and the magnetic field are the same (*longitudinal effects*), symmetry indicates that the effects are quadratically dependent on the magnetic field.
Some of the effects can be observed in single crystals, gases, electrolytes, and plasmas. In a semiconductor, the coefficients for the individual effects are dependent on the electron and hole mobilities. In an anisotropic crystal, the coefficients are also dependent on the directions of the crystal axes.
In each group the coefficients for the individual effects are dependent on one another in a simple fashion. They can be expressed either in terms of the carrier mobility or the electrical conductivity, thermal conductivity, and the Wiedemann-Franz number.
Research on these effects has contributed substantially to the understanding of current conduction in metals and later in semiconductors. In general, they occur together, and are therefore difficult to separate by experiment. As a rule, the effects are very small and so far, apart from the Hall effect and the magnetoresistance effect, they have had hardly any applications.
These effects vanish with the onset of → *superconductivity;* the conduction mechanism is different (Cooper pairs), and the magnetic field (within certain limits) has no effect on conduction (→ shielding effect, → Meissner-Ochsenfeld effect).
In superconductors of the second kind, the ef-

fects can again occur. They are then related to the flux motion in these superconductors.
Lit.: [94, 96, 100, 100, 101, 104, 110, 114]

Cushion effect
→ Poisson effect

Cyclotron resonance effect
→ Azbel-Kaner effect

DAP effect
→ liquid-crystal effects

Debye effects EM
An electric field applied to a dielectric medium aligns the permanent atomic or molecular dipoles. The frequency dependence of the dielectric constant and the dielectric loss in alternating fields can be explained phenomenologically in this way.

DEBYE, Peter Joseph Wilhelm,
Dutch-American physicist, 24 March 1884 – 2 November 1966,
Professor in Zürich, Utrecht, Göttingen, Leipzig, and Berlin (Director of the Kaiser-Wilhelm Institute for Physics), from 1940 Professor in Ithaca (New York)
Debye, P., *Phys. Z.* **13**, 97, 295 (1912)
Lit.: [72, 85] (additional lit. see Table 26)

Debye-Falkenhagen effect (1928) El
(conductivity dispersion effect)
The conductivity of an electrolytic solution is dependent on frequency. At high frequencies (a few MHz), the ion clouds cannot follow the rapid changes in the electric field, so one expects a decrease in the → relaxation effect and therefore an increase in conductivity.
This has acutally been observed by experiment (P. Debye and H. Falkenhagen 1928).

FALKENHAGEN, Hans Eduard Wilhelm, 13 May 1895 – 26 June 1971, Rostock
Debye, P. and Falkenhagen, H.E.W., *Phys. Z.* **29**, 121, 401 (1928)
Lit.: [22d] [50 Vol. 4, part 4] (additional lit. see Table 25)

Debye-Sears effect O
In a liquid standing ultrasonic waves produce a density modulation and therefore also a modulation of the refractive index. Such a liquid acts on a light beam as does a diffraction grating. The Debye-Sears effects can be used in deflecting and modulating light beams. The working liquid is usually an organic compound such as toluene.
→ Brillouin effect

SEARS, George Wallace, 9 July 1878 USA – ?
Lit.: [6c] (additional lit. see Table 39)

Déchêne effect (1935) Lu
When luminescence is stimulated by an electric field, the Déchêne effect is observed. If there is direct contact of crystals of high conductivity, field quenching can occur. The Déchêne effect is almost always linked with the → Gudden-Pohl effect and reduces this.
Lit.: [6d] (additional lit. see Table 34)

Deck effect (1964) AQ II
In the Deck effect a resonance is simulated in a multiple-particle production process in high-energy scattering experiments. For example, the resonance criterion can be met in three-pion production via the kinematics of the production of a pion together with a rho particle which decays into two pions. The true resonance, however, is the rho particle.

DECK, Robert Thomas,
American physicist, born 6 August 1935 in Philadelphia
Deck, R.T., *Phys. Rev. Letters* **13**, 169 (1964)
Lit.: [97, 100, 114]

Deflection effect
→ light deflection

Degradation effect S
The values found for the current-carrying capacity on short wire samples are normally greater than those found on coils. This is known as the degradation effect and derives from instabilities

in the conductor that ultimately lead to normal conduction. The explosive propagation of the instabilities leads to an elevated temperature in the superconductor.
Lit.: [11] (additional lit. see Table 44)

de Haas-van Alphen effect

→ Haas-van Alphen effect

Delbrück scattering

→ particle scattering

ΔE effect **Mm**

(magnetoelastic effect)
Mechanical quantities such as the elasticity are affected by a change in the magnetization of a ferromagnetic material. For example, the elastic modulus is raised by magnetization. The effect is also noticeable as an anomaly in the temperature dependence of the E modulus.
Lit.: [4, 34, 50 Vol. 4, part 4] [76] (additional lit. see Table 36)

Dember effect (1931) **PS**

(photogalvanic effect)
The occurrence of an electrical potential difference between the illuminated and unilluminated sides of a photoconductor is known as the Dember effect. The incident light is absorbed in a thin surface layer and produces electron-hole pairs there. These diffuse into the region of the photoconductor not reached by the light. If there is a difference in diffusion coefficient between the electrons and holes, there is partial charge separation, giving rise to an electric field. This retards the more rapidly diffusing charge carriers and accelerates the slower ones. This in turn leads to an equilibrium field in which the two types of charge carrier move equally rapidly. The corresponding voltage is called the Dember voltage. The effect has been used, for example, to measure the transport of light-induced holes in AgBr (I) crystals.
It also appears possible to measure material purity if the material is exposed to a laser beam, and, in addition to the Dember voltage, the relaxation time and temperature dependence of the diffusion current are measured.

DEMBER, Harry,
German physicist, 11 July 1882 Leimbach – 22 March 1943 New York

Dember, H., *Fortschr. Mineral* **16**, 4 (1931)
Dember, H., *Phys. Z.* **32**, 554; **33**, 207 (1931)
Kahan, A.M., *Photogr. Sci. Eng.* **21**, 237 (1977)
Berman, A., Dickson, C.R., and Zare, R.N.: *Proceedings of the Technical Program, Electron-Optical Systems Design Conference*. New York, 14 – 16 September 1976
Lit.: [6d, 74, 75, 78] (additional lit. see Table 42)

Destriau effect (1936) **Lu**

(alternating field luminescence)
In research on the → Déchêne effect, Destriau discovered that powdered Cu-doped ZnS/ZnO phospors luminesce on the application of an alternating potential of between 50 and 500 V. To avoid gas discharges, it is recommended that the substances should be embedded in an insulator. The effect is observed primarily in the edge or surface layers.
The luminescence section thickness increases exponentially with the applied voltage. The effect has been used in flat displays for panels and scales and as emergency lighting sources.

Destriau, G., *J. Chim. Phys.* **33**, 587 (1936)
Shaposhnikov, A.N., *Meas. Tech. USSR* **4**, 585 (1974)
Theis, D., *J. of Luminescence* **23**, 1, 2, 191 (1981)
Lit.: [6d, 72 – 75, 85]

Diffraction **G**

Diffraction means deviation from a straight-line propagation in light or in waves generally. This is to be distinguished from reflection, refraction, and scattering. Diffraction is observed whenever free propagation is affected by an obstacle (diaphragm, screen, or body). Diffraction resembles interference in being characteristic of any type of wave motion. Diffraction effects can be explained on the basis of *Huygens' principle*, according to which any point on a wavefront is a starting point for an elementary wave. The superposition of all the elementary waves gives the new wavefront. Any perturbation in the wavefront gives rise to an interference pattern, the diffraction pattern, by superposition of the elementary waves arising from different points on the obstacle.
Diffraction limits the resolution of optical in-

struments such as the microscope, telescope, and spectrometer.

Diffraction effects can be divided into two groups: Fraunhofer diffraction and Fresnel diffraction.

Fraunhofer diffraction is observed if the size of the diffracting aperture is small relative to the distance to the observation point. Fraunhofer diffraction can be reproduced by experiment by placing the light source at the focus of a lens. A second lens is used to bring the parallel light together at an observation screen.

Fresnel diffraction occurs when the size of the diffracting object is comparable with the distances from source to object and from object to observation point. Divergent light beams are used in Fresnel diffraction. The observed light beams have different deflection angles.

FRAUNHOFER, Joseph von,
German physicist, 6 March 1787 – 7 June 1826, from 1823 Professor in Munich

FRESNEL, Augustin Jean,
French physicist, 10 May 1788 – 14 July 1827 Ville d'Avray

Lit.: [6c, 9, 16, 22c, 38 Vol. II, Vol. VIII, 50 Vol. II, 1] [61 Vol. 4] (additional lit. see Table 19)

Diffraction filter-beam focussing effect (1968)

This effect can be used to raise the transmission of a mechanical diffraction filter for short-wave light while reducing that for long-wave light by several orders of magnitude.

A diffraction filter consists of a row of slits aligned behind one another and employs the successive diffraction at the slits to remove most of the long-wave radiation from the beam. Arrangement of the slits next to and behind one another leads to grating-type Soller apertures, which are used as mechanical collimators in the soft X-ray region and in the far ultraviolet (XUV). When the latter are properly dimensioned, they also act as diffraction filters. If, in addition, identical gratings are inserted in these diffraction filters (such a grating is an arrangement of mechanical slit apertures), the long-wave radiation can be further filtered out by diffraction at the slit apertures. On the other hand, short-wave radiation exhibits an intensity increase because of the formation of slit patterns at the slits of the subsequent grating. One may therefore speak of beam focussing for short-wave light: improved transmission is observed for short wavelengths, while for long wavelengths transmission is reduced.

An appropriate aperture geometry (slit width, separation of the successive slits and/or number of successive gratings) enables one to shift the transmission region from the XUV via the visible region to the IR.

(→ see also Table 39)

Schmidtke, G., *Z. Angew. Phys.* **25**, 314 (1968)
Schmidtke, G., *Optik* **28**, 578 (1969)
Schmidtke, G., *Appl. Optics* **9**, 447 (1970)

Lit.: [96, 98, 100, 106]

Diffusion G

This is a transport process in which particles move without an external force: e.g., a spatial concentration gradient can produce particle movement. Examples from everyday life are the propagation of odors and the mixing of liquids of different color or sweetness. A consequence of *Brownian movement* is that more particles move from a region of higher concentration to one of lower concentration than do the converse. A very much smaller drift or diffusion velocity is superimposed on the thermal velocity.

Diffusion occurs in gases, liquids, and solids. The diffusion rate decreases sharply in that sequence. Also, in solids it is very much dependent on various diffusion mechanisms.

The diffusion of foreign atoms in solids plays an important part, for example, in semi-conductor doping, surface treatment, and corrosion.

The dependence of the diffusion coefficient on particle mass has been used to separate gases. These methods are used to separate isotopes from one another. An example of considerable economic importance is the enrichment of 235 UF 6 in natural uranium hexafluoride, which consists of 99,3 % 238 UF 6. There are various effects related to diffusion, such as *thermal diffusion* and *pressure diffusion*, → diffusion thermo-effects (*Dufour effect, Ludwig-Soret effect*), → Knudsen effect (*thermal effusion*).

The diffusion process can also be derived from transport theory.

→ cross effects, → transport effects

Lit.: [6a, 22a, 50 Vol. III, 72 – 77] (additional lit. see Table 19)

Diffusion aftereffect

→ magnetic aftereffects

Diffusion thermo-effects TK

(1856, 1872, 1879)
(Dufour effect, Ludwig-Soret effect)
A substance dissolved in a solvent should behave like a low-density gas with respect to its equation of state. This applies to dilute solutions, the gas pressure being replaced by the osmotic pressure. Temperature differences within the solution then lead to concentration differences.
Ludwig and later Soret showed that concentration differences arise in a solution containing temperature differences. This is known as the *Ludwig-Soret effect* (1856/1879). Dufour observed the inverse effect: the diffusion of two substances in a solution produces a temperature gradient (*Dufour effect* 1872).
The relationships are as follows:
Temperature gradient → current flow (thermal diffusion flux): Ludwig-Soret effect
Concentration gradient → heat flux (temperature gradient): Dufour effect

DUFOUR, Louis,
Swiss physicist, 17 February 1832 – 1892

LUDWIG, Karl Friedrich Wilhelm, 29 December 1816 – 23 April 1895

SORET, Jacques, 30 June 1827 – ?
Ludwig, K.F.W., *Wien. Ber.* **20**, 539 (1856)
Soret, J., *Arch. Sci. Phys. Nat.* (3) **2**, 48 (1879)
Lit.: [23, 50 Vol. III, part 1, section 2, 70] (additional lit. see Table 54)

Dipole effects AQ I

(direction effect, induction effect)
The forces occurring between valency-saturated molecules are called *van der Waals forces*. These are produced by the dipole or *direction effect*, the induction effect, and the → dispersion effect.
In the dipole effect, there is an interaction between two molecules with permanent dipoles.
At high temperatures, allowance must be made for the fact that the molecules are deformed and that additional electric moments may be induced (*induction effect*).
Lit.: [75] (additional lit. see Table 20)

Direction effect

→ dipole effect

Dispersion effects G

Dispersion of waves leads to a dependence of the refractive index on the wavelength or frequency. The phase velocity is also dependent on the wavelength. Depending on whether the change in refractive index with wavelength is positive or negative, *normal* or *anomalous dispersion* results. The frequency dependence of the material constants gives rise to various effects which influence, e.g., the electrooptic and magnetooptic effects. Dispersion also plays an important part in optical communications, since it affects the shapes of light pulses propagating in glass fibers.
Lit.: [72, 76] (additional lit. see Table 19)

Dispersion effects AQ I

(intermolecular interaction)
In a molecular assembly, electron movement leads to the production of fluctuating dipoles, which polarize neighboring molecules. The resulting forces are called dispersion forces, and these make up the main contribution to intermolecular interaction.
→ Dipole effects
Lit.: [74] (additional lit. see Table 20)

Dispersion effect in conductivity

→ Debye-Falkenhagen effect

Displacement effect

→ Meissner effect

Dissociation-voltage effect El

(1931)
(M. Wien and L. Onsager 1934)
A strong electric field applied to a weak electrolyte can produce a change in its dissociation. The effect may be 5 to 10 times larger than the normal voltage effect (→ Wien effect). One consequently finds a corresponding increase in the conductivity.
The electric field produces additional dissociation in otherwise undissociated molecules, which causes an increase in the conductivity.

ONSAGER, Lars,
Norwegian-American physicist, 27 November 1903 – 5 October 1976, Professor in Baltimore, Providence, and New Haven
Lit.: [22d] (additional lit. see Table 25)

Dissymmetry effect

→ galvanomagnetic effects

Doppler effect, acoustic **Me**

(1842)

If a source of sound is in motion relative to an observer, there is a rise in the frequency recorded if the body approaches the observer and a reduction if it recedes from the observer. This occurs because, depending on the direction of motion, the observer receives a greater or smaller number of waves per unit time. The acoustic Doppler effect was observed in 1845 by Buys-Ballot by means of moving trains. As the observer and source (or sender) can be moving independently of one another, the observable frequency change is dependent on the speeds with which the source and recipient are moving. In fact, it is found that the effect is dependent only on the relative velocity.

→ optical Doppler effect

DOPPLER, Christian,
Austrian physicist, 29 November 1803 – 17 March 1853, Professor in Prague and from 1850 in Vienna

Doppler, Chr., *Abh. Kgl. Böhm. Ges.* III (1845)
Doppler, Chr., *Böhm Ges. Wiss. N.F.* **5**, 419 (1845)

Lit.: *textbooks, dictionaries*

Doppler effect, optical

→ optical Doppler effect

Dorn effect (1878) **Ek**

This is also called the *sedimentation potential.* If charged solid particles fall through a fluid, e.g., on account of their gravitational potential, a potential difference arises between the surface of the fluid and the bottom which inhibits the particle motion.

The converse is the *flow potential.* When a liquid flows through a capillary, a potential difference arises between its ends.

DORN, Friedrich Ernst,
German physicist, 27 July 1848 – 13 June 1916

Dorn, F.E., *Annalen d. Physik u. Chemie* **5**, 20 (1878)
Dorn, F.E., *Annalen d. Physik u. Chemie* **9**, 513 (1880)

Lit.: [12, 75, 77]

Drag effect **SC**

(photon drag effect, phonon drag effect)

The drag effect is a transport effect derived from the nonequilibrium phonon distribution in a semiconductor which influences the charge carrier distribution. For example, a temperature gradient in a semiconductor produces a phonon excess on the hotter side which the charge carriers sense. Then, under certain conditions (at low temperatures), an appreciable increase in the therme-EMF may be observed.

Allowance must be made for the drag effect in all transport phenomena (thermoelectric and thermomagnetic phenomena).

Lit.: [2, 20, 75] (additional lit. see Table 48)

Drift inhibition

→ braking effects

Dufour effect

→ diffusion thermo effects

Dynamoelectric principle (1866/1867) **EM**

(Werner von Siemens)

Werner von Siemens replaced the permanent magnet in a DC generator by an electromagnet which was excited by the armature voltage (*self-excitation*). He showed that the residual magnetism in the iron circuit of the generator was sufficient to induce a small voltage, the remanence voltage. This produces a small current which passes through the armature and through the parallel exciting winding. The current strengthens the magnetism, so the current increases, and so on. The magnetic fields thus produced are limited only by the saturation of the iron used. The dynamo effect forms the basis for all high-current engineering. It was also discovered simultaneously by several other researchers (for example, Wheatstone), but Werner von Siemens recognized the significance of the principle first and described the possibilities it offered to make energy available wherever it was needed.

SIEMENS, Werner von,
German inventor and manufacturer,
13 December 1816 Lenthe – 6 December 1892 Berlin, 1873 Member of the Prussian Academy of Sciences

Siemens, W. v., *Sitzungsber. Dtsch. Akad. Wiss. Berlin*, 17 January 1867
Lit.: *textbooks, dictionaries*, [50 Vol. 4, 1] (additional lit. see Table 26)

Early effect (1953) SCC

This effect in bipolar transistors involves a change in base width as a function of base-collector bias voltage. The base-collector barrier layer extends in accordance with the bias voltage applied. In particular, the Early effect influences the output admittance and the feedback admittance of a transistor.
→ transistor effect
Early, J.M., *Proc. IRE* **40**, 1401 (1953)
Early, J.M., *Bell Syst. Tech. J.* **32**, 517 (1954)
Malhi, S.D. and Salama, C.A., *Solid State Electron* **21**, 9, 1187 (1978)
Lit.: [63, 84 Vol. III] (additional lit. see Table 49)

Eddy-current effects

→ Arago's experiment, → skin effect

Edison effect (1885)

Edison was the first to suspect that negative electricity was emitted by filaments heated to high temperatures. The phenomenon was observed by W.H. Prece in 1885 and was called the Edison effect.
→ Richardson effect
Prece, W.H., *Proc. Roy. Soc.* **38**, 219 (1885)
Lit.: [50 Vol. IV, 4] [86 Vol. XIII, Vol. XIV, 87 Vol. XXI]

Effects in electrolytes El

Electrolytes are substances that dissociate into ions in the molten state or in solution (generally aqueous). Their electrical conductivity is governed by the ionic mobility. The following groups of experiments give an indication of the processes in electrolytes: investigations into processes at the electrodes (for example, Faraday's laws), into processes within the electrolytes (dissociation), and into special effects such as, for example, → the braking effects, → the Debye effects, energy consumption or production in electrochemical processes, and → electrokinetic effects.
Lit.: [6b, 6d, 22b, 50 Vol. 4, part 4] (additional lit. see Table 25)

Effusion

→ Knudsen effect

Einstein effects R

Three important effects called Einstein effects can be derived from the general theory of relativity. These are the → perihelion advance of the planets, → light deflection in the gravitational field of a massive star, and → gravitational frequency shift in spectral lines. The effects are derived by solving Einstein's field equations.

EINSTEIN, Albert,
German-American physicist, 14 March 1879 – 18 April 1955
Lit.: [6c, 10, 59] (additional lit. see Table 43)

Einstein-de Haas effect (1915) EM

(converse of the → Barnett effect, predicted by O.W. Richardson in 1908)
A freely suspended iron rod is set into rotation by sudden magnetization. This effect is explained by the conservation of the total angular momentum, as is the Barnett effect.
Experiment shows that the magnetic moment, for example, of an iron atom, is produced by spinning and circulating electrons. The effect is actually very small.

EINSTEIN, Albert,
German-American physicist, 14 March 1879 – 18 April 1955

HAAS, Wander Johannes de,
Dutch physicist, 2 March 1878 – 26 April 1960
Einstein, A., *Verh. Dtsch. Phys. Ges.* **17**, 152 (1915)
Einstein, A., *Verh. Dtsch. Phys. Ges.* **18**, 173 (1916)
de Haas, W.J., *Verh. Dtsch. Phys. Ges.* **18**, 173 (1916)
Barnett, S.J., *Rev. Mod. Phys.* **7**, 129 (1935)
Lit.: [6b, 17, 50 Vol. 4, part 4] (additional lit. see Table 26)

Elastic limit effect

→ Portevin-le Chatelier effect

Elastic shape-memory effect

→ memory effect

Electrical birefringence

→ electrooptic effects

Electrocaloric effect LT

This is an adiabatic cooling at low temperatures similar to the *magnetocaloric effect*. Use is made of the fact that the change in a quantity is dependent on the system entropy. The molecular dipoles of mixed ionic crystals are aligned by pressure or by means of an electric field. At low temperatures, the field is removed, and the molecular directions thus gain two additional degrees of freedom. This corresponds to a temperature reduction of a few degrees.
The effect can be used in producing low temperatures. The converse of the electrocaloric effect is the → pyroelectric effect.
Lit.: [3, 6d, 75] (additional lit. see Table 33)

Electrocapillarity (1875) Ek

The mechanical surface tension of a liquid is dependent on its charge. This phenomenon is used in electrocapillarity. The change in height of the meniscus of a mercury column in a capillary produced by an electrical voltage is a measure of the voltage. The capillary electrometer has virtually no significance nowadays.
Lippmann, G., *Ann. Chim. Phys.* (5) **5**, 594 (1875)
Lit.: [6b, 22b, 50 Vol. 4, part 4, 69, 77] (additional lit. see Table 24)

Electroendosmosis

→ electrokinetic effects, → electroosmosis

Electrohydrodynamic effects

→ liquid-crystal effects

Electrokinetic effects Ek

It is possible for one phase to move relative to the other at the boundary of a two-phase electrochemical system due to an applied voltage. An electrical potential can arise from the motion of the phases relative to one another. The following effects are distinguished:
→ electrophoresis (cataphoresis, anaphoresis), → electroosmosis (electroendosmosis), → flow current (diaphragm current), and flow potential, → Dorn effect, → electrocapillarity, → electroviscous effect.

Table 3. Electrokinetic effects

	An electric current moves	Name
Within walls at rest	a fluid	Electroosmosis
Within a fluid at rest	charged solid particles	Electrophoresis
	An electric current is produced by the motion of	Name
Within rigid walls at rest	a fluid	Flow current, diaphragm current
Within a fluid at rest	charged solid particles	Cataphoretic currents, Dorn effect

Lit.: [6b, 22d, 24, 50 Vol. 4, part 4] (additional lit. see Table 24)

Electrolytes

→ Effects in electrolytes

Electronic Kerr effect

→ Kerr effect

Electrooptic birefringence

→ Kerr effect

Electrooptic effects Eo

The electrooptic (and magnetooptic) effects may be divided into two major groups (see Table 41): 1. Changes occur in the optical parameters of a material produced by external fields. 2. Changes occur in the emission from light sources in external fields.
1. Light travelling through an optically transparent material is affected by electric fields. Two experimental systems are conceivable: observation in the field direction and perpendicular to it. In a solid or liquid, the refractive index changes, and birefringence or rotation of the plane of polarization results.
2. Light-emitting atoms show a characteristic line splitting in an electric field. This can be very complicated, depending on the number of shell electrons involved.

Table 4. Electrooptic Effects

Occurrence / Produced by an electric field	Change in refractive index or polarization direction	Line splitting
Transverse to the propagation direction	Kerr effect (electrooptic Kerr effect), Neugebauer effect	Stark effect Linear Stark effekt Quadratic Stark effect
Longitudinal to the propagation direction	Pockels effect, linear birefringence	

Lit.: [6c, 18, 22c, 50 Vol. II, 2.2] (additional lit. see Table 27)

Electroosmosis **Ek**

(electroendosmosis)

An electric field applied to charged particles suspended in a liquid causes them to move through a diaphragm. This effect is used, for example, in dewatering peat, and in purifying silica or alumina.

Lit.: [24, 50 Vol. 4, part 4] (additional lit. see Table 24)

Electrophoresis **Ek**

(Reuss 1807, Wiedemann 1852; Quincke 1861)
(cataphoresis, anaphoresis)

Electrophoresis means the motion of electrically charged particles suspended in a fluid or colloidal particles in an electric field. Electrophoresis is very important in analytical chemistry and in medicine. The converse of electrophoresis is denoted by the *electrophoretic potential*, and the currents arising in this way are called *electrophoretic currents*.

→ Dorn effect

QUINCKE, Georg Hermann,
German physicist, 19 November 1834 – 13 January 1924

Lit.: [24, 50 Vol. 4, part 4] (additional lit. see Table 28)

Electrophoretic currents

→ electrophoresis

Electrophoretic effect

→ slowing down effect

Electrophoretic potential

→ electrophoresis

Electrophotoluminescence

→ luminescence effects

Electrostriction **ET**

An electric field acting on an insulator can produce changes in shape and volume or elastic strains (electrostriction) in the material. These changes are not the same as those caused by condenser forces. In crystals with a center of symmetry, the deformation is proportional to the square of the potential and is thus dependent on the square of the polarization, while being independent of the direction. If an alternating electric field is superimposed, linear and quadratic alternating-field electrostriction effects are observed, which vanish above a frequency characteristic of the material (relaxation frequency), since the dipoles in the material can no longer follow the changes in the field.

Electrostriction has also been observed in liquids. Here, under certain conditions, electrostriction leads to a volume decrease, because the dipoles in the liquid are compressed.

Quadratic electrostriction has little technical significance.

→ piezoelectric effect

Guntersdorfer, M. *Phys. Unserer Zeit* **7**, 48 (1976)

Lit.: [6b, 18, 22b, 50 Vol. 4, part 4] (additional lit. see Table 24)

Electrothermal inhomogeneous effect

→ thermoelectric effects, → current conduction effects

Electro-tunnelling effect (1960) **S**

In superconductors separated by an insulator of thickness less than 10^{-9} m, single electrons can tunnel from one superconductor to the other. This is called the electron-tunnelling effect. A nonlinear current-voltage characteristic is observed when one or both of the metals becomes superconducting.

→ Josephson effects

Giaever, I., *Funkschau* (3) **33**, 107 (1961)
Giaever, I. and Megerle, K., *IRE Trans Electron Devices,* ED **9**, 469 (1962)
Lit.: [11, 15, 44, 75] (additional lit. see Table 44)

Electroviscous effect Ek

By this is meant the change in viscosity for certain colloids when an electrolyte is added. A fluid flowing in a capillary produces a flow potential, which is responsible for electroosmotic fluid movement in the opposite direction. The consequence is an apparent increase in the viscosity.
Lit.: [75] (additional lit. see Table 24)

Elia's fire

→ corona effect

St. Elmo's fire

→ corona effect

Emitter-dip effect (1968) SCC

This is an interference effect which occurs during diffusion into the base and emitter regions of a transistor as the dope concentration within the chip decreases. This restricts the minimal base width in a high-frequency transistor and therefore the transistor frequency limit. During emitter doping, the doping front shifts that in the base further on.
Willoughby, A.F.W., *J. Mater, Sci.* **3**, 89 (1968)
Lit.: [84 Vol. III] (additional lit. see Table 49)

Energy effect

→ red shift

Eötvös effect (1919) Me

In a system moving relative to the Earth's surface, the *Coriolis force* causes a change in its acceleration due to gravity. If a body is moving from west (east) to east (west), it rotates more rapidly (less rapidly) than the Earth. On account of its larger (smaller) centrifugal acceleration, it appears to be lighter (heavier).
The terrestrial acceleration is then smaller (greater).
A body moving at ± 60 km.h^{-1} and of mass 100 kg, experiences a change in weight of ∓ 15 g.
Eötvös observed this effect with a balance rotating very rapidly about its point of suspension. A deflection of the balance beam is observed.
The effect is large when resonance occurs.
Eötvös, Loránd Baron von,
Hungarian physicist, 27 July 1848 – 8 April 1919, from 1872 Professor in Budapest
Eötvös, L. *Ann. Phys.* (Leipzig) **59**, 10 (1919)
Eötvös, L., *Ann. Phys.* (Leipzig) **68**, 56 (1922)
Lit.: [6a, 22a, 69, 73, 76] (additional lit. see Table 35)

Eötvös experiment (1909) R

In 1909, Eötvös performed the following experiment, in order to demonstrate the equivalence of gravitational and inertial masses. Materials differing in density were attached to a torsion balance. The tangential components of the gravitational force and the centrifugal force act on the gravitational or inertial masses. If gravitational and inertial masses are different, there should be a tilt in the balance, or a twist in the filament on which the balance is suspended. The measurement uncertainty in the apparatus was less than 10^{-9}. No difference could be detected between gravitational and inertial masses.
Lit.: [6a, 22a, 69, 73, 76] (additional lit. see Table 43)

Ettingshausen effect

→ galvanomagnetic effects

Ettingshausen-Nernst effect

→ thermomagnetic effects

Ettingshausen-Nernst effect, second

→ thermomagnetic effects

Exchange effects AQ I

These are all effects related to the exchange of quantum properties between indistinguishable particles. Examples are the increase or decrease in Coulomb energy and the consequent shift in energy levels in atoms. This effect is derived from the antisymmetry in the wave-function on particle exchange.
Lit.: [38 Vol. III, IV, IX, 72] (additional lit. see Table 20)

F effect

→ substituent effect

Fano effect (1969) AQ II

Photoelectrons from alkali atoms, produced upon absorption of cirularly polarized light, exhibit spin polarization. The spins of the absorbed photons are either parallel or antiparallel to the direction of beam propagation.
The effect is related to the dependence of the photoionization cross section on the electron spin direction. The effect is used to produce polarized electrons.

FANO, Ugo,
Italian-American physicist, 28 July 1912 Turin
Fano, U., *Phys. Rev.* **178,** 131 (1969)
Kessler, J., *Phys. Blätter* **27,** 161 (1971)
Kessler, J., *Phys. Blätter* **38,** 31 (1982)
Kessler, J., *Phys. Unserer Zeit* **6,** 40 (1975)
Lit.: [75] (additional lit. see Table 21)

Faraday effect (1845) Mo

(magnetic rotation of the plane of polarization)
In an optically inactive material linearly polarized electromagnetic waves cause rotation of the plane of polarization when a magnetic field is applied parallel to the propagation direction. The angle of rotation is proportional to the lenght of substance traversed and to the magnetic field strength.
The coefficient of proportionality is named after Verdet, who, with Wiedemann, measured the effect precisely.
The effect is used in modern optics in connection with lasers and also in the microwave region, e.g., in modulators.
In contrast to the → Cotton-Mouton effect and to optical activity, the Faraday effect is not reciprocal: if the light emerging from the substance is reflected back into it, the angle of rotation is doubled.

FARADAY, Michael,
British physicist and chemist,
22 September 1791 – 25 August 1867
WIEDEMANN, Gustav Heinrich,
2 October 1826 Berlin – 23 March 1899 Leipzig
VERDET, Marc Emile.
13 March 1824 – 3 June 1866
Faraday, M., *Philos. Mag.* (3) **29,** 153 (1846)
Faraday, M., *Philos. Trans. R. Soc. London* **1,** 1 (1846)
Faraday, M. *Pogg. Ann.* **68,** 105 (1846)
Wiedemann, G., *Pogg. Ann.* **82,** 215 (1851)
Verdet, E., *Ann. Chim. Phys.* **41,** 570 (1854)
Lit.: [3, 6c, 9, 22c, 26, 50 Vol. II, 2.2] (additional lit. see Table 37)

Ferroelectric Barkhausen effect EM

Ferroelectric materials exhibit a process analogous to the magnetic → Barkhausen effect. If the polarization direction is changed at low electric field strengths, switching of ferroelectric domains can be observed. A sharp change is experienced as noise. The statistical reversal of ferroelectric domains (polarization directions) is linked to the production of charge pulses, which can be detected by sensitive measuring equipment.
The effect has virtually no practical significance. It occurs, for example, in gadolinium molybdate.
Lit.: [77] (additional lit. see Table 26)

Field effect (1926) EM

(field electron emission)
This is the emission of electrons under high vacuum from fine metal points upon application of a high voltage. The effect is used in the field-electron microscope, in which the point is imaged on a fluorescent screen. The crystal structure is then visible.
Rother, F., *Ann. Phys.* (Leipzig) **81,** 317 (1926)
de Bruyne, N.A., *Phys. Rev.* **35,** 172 (1930)
Lit.: [50 Vol. 4, part 4] (additional lit. see Table 26)

Field effect SC

An electric field produces a local change in the charge carrier concentration in a semiconductor. This effect is used in field-effect transistors and in MOS devices for resistors and diodes.
Lit.: [53] (additional lit. see Table 48)

Field effect

→ liquid-crystal effects

Field-strength effect

→ Wien effect

Fizeau's experiment (1851) O

(Fizeau effect)
In 1818 Fresnel predicted that the speed of light and the speed of the body in which the light propagates may not simply be added. This was

experimentally confirmed in 1851 by Fizeau, who split a parallel beam into two partial beams passing through two tubes in which water flowed at high speed, for example, in a clockwise direction. The light leaving the end of the tube was reflected in such a way that the light from tube 1 entered tube 2 and vice versa. This arrangement therefore constitutes a ring interferometer. The shift of the interference fringes is then observed. The speed of light in the body can be calculated by means of *Fresnel's transport coefficient*. The shift calculated from the transport coefficient agrees with the measured value within the error limits.
The transport coefficient may be derived from relativity theory, and it is found to represent a first approximation, which in general is sufficient. Lorentz also pointed out that the wavelength of light in a moving body must alter (the Doppler effect). This is related to a change in the refractive index, a fact which was detected by Zeeman, and which also follows from relativity theory.

FIZEAU, Armand Hippolyte Louis,
French physicist, 23 September 1819 – 18 September 1896 Venteuil
Fizeau, A.H., *Ann. Phys.* **82,** 124 (1851)
Fizeau, A.H., *C.R.,* **33,** 349 (1851)
Buchwald, E., *Naturwissenschaften* **22,** 519 (1951)
Lit.: [6c, 22c, 55, 61 Vol. 4, 72 – 78] (additional lit. see Table 39)

Flow current (1859) Ek

(Quincke)
(flow potential)
If a liquid passes through a diaphragm, it acquires a charge opposite to that of the diaphragm and thus produces a flow-dependent voltage or a potential generated by the current flow.
Lit.: [50 Vol. 4, part 4] [75, 77] (additional lit. see Table 23)

Flow voltage

→ flow current

Fluctuation effects

→ shot effect

Foldy effect (1951) AQ II

A neutron has a nonvanishing charge distribution in its interior which is responsible for the anomalous magnetic moment of a neutron. The neutron performs oscillations about its mean position, and produces an electric field which can interact with an electron.
This effect has to be taken into account in relativistic theory and gives rise to so-called *Foldy-terms*.
Lit.: [75] (additional lit. see Table 21)

Foucault's experiment (1850) Me

The plane of an oscillating pendulum does not remain fixed in space because of the Coriolis force. Viriani observed this effect in 1661. In 1850, Foucault detected the Earth's rotation in Paris, where he used a pendulum of length 67 m and a bob of mass 28 kg. The path of the pendulum describes a rosette. The rotation of the plane is dependent on the geographic latitude and, for example, at Berlin is about 12° an hour.

FOUCAULT, Jean Bernard Léon,
French physicist and astronomer, 18 September 1819 – 11 February 1868 Paris
Ann. Chim. Phys. Ser. III T XXXV (1852)
Lit.: [6a, 13a, 27 Vol. I] (additional lit. see Table 35)

Fountain effect

→ Onnes effect

Fractional quantum Hall effect (1984) Ga

By analogy with the → von-Klitzing-effect (also called the integer quantum Hall effect), the Hall voltage of two-dimensional electron systems shows anomalies that indicate that a stable ground state of the electron system can also exist if the proportion ν, from the number of magnetic flux quanta per cm^2 and the number of electrons per cm^2, is rational. This so-called fractional quantum Hall effect occurs only for semiconductor samples with large electron mobility ($\mu = 10^6$ cm^2/Vs). A prerequisite for the effect is the establishment of a two-dimensional electron gas, e.g. in inversion layers on semiconductor surfaces or in thin semiconductor layers both at low temperature (≤ 1 K). Up until now this has

only been possible with silicon field-effect transistors or GaAS-(AlGa)As heterostructures. The effect, i.e. the establishment of a multi-particle ground state, appears when these samples are placed in a strong magnetic field ($\geq$ 10 T). So far rational ν with odd denominator (e.g. ν = 1/3, 2/3, 4/3, 5/3, ..., 2/5, 3/5, ..., 2/7, 3/7, 4/7, ...) and also even denominator (e.g. ν = 5/2, 7/2, 9/4, 11/4, ...) have been reported.
Stoermer, H.L., *Festkörperprobleme* **XXIV**, Vieweg, Brunswick (1984), p. 25.

Franck-Condon principle (1926) AQ II

The Franck-Condon principle states that optically stimulated electronic transitions in molecules or crystals are fast by comparison with changes in nuclear separation or momentum, which lead to changes in internal potential energy. The vibrational transitions in the molecular bands associated with the electron jump occur preferentially at the turning points of the vibration. The electrons reach the final state again at a turning point of the vibration. The electron rearrangement is so rapid that the nuclear positions remain virtually unchanged. If the vibrational frequencies of the initial and final states in the transition are plotted in a rectangular coordinate system, the *Franck-Condon parabolas* result. The width of the parabolas is controlled by the change in binding energy associated with the transition.

FRANCK, James,
German physicist, 26 August 1882 – 21 May 1964,
from 1916 Professor in Berlin, 1920 – 1933 in Göttingen, 1938 – 1947 in Chicago

CONDON, Edward Uhler,
American physicist, 2 March 1902 – 26 March 1974,
Professor in Princeton and Saint Louis

Franck, J., *Z. Phys. Chem.* **120**, 144 (1926)
Condon, E.U., *Phys. Rev.* (62) **32**, 858 (1928)
Lit.: [6d, 7, 17, 22d, 67, 70 Vol. II, 90] (additional lit. see Table 21)

Franck-Hertz experiment (1913) AQ II

Franck and Hertz were able to show from experiments on collisional electron excitation that atoms possess discrete energy levels. Electrons were accelerated in a tube filled with mercury vapor. At first, the current increased in proportion to the voltage. The electrons could reach an acceptor electrode with a small negative bias through the acceleration electrode. At a certain voltage, the current suddenly decreased: the electron energy had become sufficiently large for them to be able to surrender energy to the mercury atoms by inelastic collision. They therefore no longer reached the acceptor electrode. The first excitation level lies at 4.9 eV, and the associated emission at λ = 253.7 nm can be observed directly. The repetition of maxima and minima on the Franck-Hertz curve is due to repeated excitation of the same transition at different points in collisional space. With a suitable electrode arrangement, further resonances can be observed at integer multiples of 4.9 eV.
The experiment provided substantial support for Bohr's theory.

FRANCK: → Franck-Condon principle

HERTZ, Gustav,
German physicist, 22 July 1887 – 30 October 1975 Berlin, from 1925 Professor in Halle, 1928 – 1944 in Berlin, 1945 – 1953 in the Soviet Union (Sukhumi), from 1954 Professor in Leipzig

Franck, J. and Hertz, G., *Verh. Dtsch. Phys. Ges.* **15**, 34 (1913)
Lit.: [6d, 7, 17, 22d, 41, 66, 70 Vol. II, 79, 90] (additional lit. see Table 21)

Franz-Keldysh effect (1958) SC

(electroabsorption and reflection)
An electric field applied to a semiconductor or insulator shifts the optical absorption edge to smaller photon energies. In an electric field, there is a finite probability that the electrons are located in the forbidden energy band near the band edge. Consequently, the absorption at a given photon energy becomes stronger as the field strength increases in this region. This can be interpreted as a shift in the absorption edge. For example, a field of 5 x 10^4 V/m produces a shift of about 10 meV. Reflection is also affected. Because of its short lag the effect can be used in modulating light up to high frequencies. It also affects the behavior of infrared detectors.
Franz, W., *Z. f. Naturf.* **13a**, 484 (1958)
Keldysh, L.V., *Soviet Phys. JETP* **34**, 665 (1958)
Chambuleyron, I.E., *Solid State Electron* **7**, 605 (1976)
Lit.: [6d, 75] (additional lit. see Table 48)

Fraunhofer diffraction
→ diffraction

Fresnel diffraction
→ diffraction

Funkel effect
In an oxide cathode, or in any cathode that does not consist of pure metal, there can be sudden changes in work function during operation due to changes in the cathode surface. These fluctuations in work function produce fluctuations in the anode current, which, on account of their statistical distribution, are detected as noise. The Funkel effect can exceed the normal → shot effect by several orders of magnitude. Its spectrum lies approximately in the region below 1 kHz. Similar anode-current fluctuations in a valve can occur because of positive ions entering the space-charge region in front of the cathode. This is called the *anomalous Funkel effect* and occurs in all cathodes, particularly those of tungsten and molybdenum.
→ Schottky effect
Rothe, H. and Kleen, W.: *Electron Tubes as Input Amplifiers* [in German], Becker & Erler, Berlin (1940)
Bartels, H.: *Basis of Amplifier Techniques* [in German], 2nd Ed.. S. Hirzel, Leipzig 1944
Lit.: [76, 85]

Galvanomagnetic effects Ga
This term covers effects that occur in a conductor in a homogeneous magnetic field carrying only one current: the Hall effect, the anomalous Hall effect, the Corbino effect, the dissymmetry effect, the Ettingshausen effect, change in resistance in a magnetic field (the Thomson effect), and the Peltier effect between magnetized and unmagnetized materials (the Nernst effect).
Lit.: [1, 6b, 20, 22d, 30, 38 Vol. VIII, 50 Vol. IV, 4] [53, 62, 71] (additional lit. see Table 30)

Corbino effect (1911)
The Corbino effect is a combination of the → Hall effect and the magnetic resistance change in a conductor. An electric current is supplied to the midpoint of a foil and tapped off along the edge. A magnetic field is switched on and off perpendicular to the foil. Induction currents are observed in a coil concentric with the foil when the magnetic field is switched. If the foil and magnetic field are not perpendicular to one another, the foil tends to set itself perpendicular to it. The spiral paths of the electrons cause the resistance of the material to increase.
The Corbino effect is in fact identical with the Hall effect and the magnetoresistance effect, and was formerly used to measure the constants of these effects. Today, the effect is used to measure the charge carrier mobility in semiconductors.
Also, Corbino foils made of indium antimonide provide high-performance switches (200 to 1000 A). In an axial magnetic field the resistance change is by a factor of about 200 at medium currents (20 A).

CORBINO, Orso Mario,
Italian physicist, 30 April 1876 – 23 January 1937
Corbino, O.M., *Phys. Z.* **12,** 561 (1911)
Lit.: [6b, 72, 77]

Dissymmetry effect (1916)
If the magnetic field in the Hall effect is reversed, the sign of the Hall voltage is also reversed, but takes another value. The difference in the two values is proportional to the square of the field strength. The longitudinal change in resistance additive with the Hall effect remains constant on reversal. Therefore, the Hall voltage must vary, depending on the direction of the magnetic field.
Wold, J., *Phys. Rev.* **7,** 169 (1916)
Lit.: [50 Vol. 4, part 4]

Ettingshausen effect
A homogeneous conductor carrying direct current and being located in a magnetic field exhibits not only the → Hall effect but also a temperature gradient, which is proportional to the current and the magnetic field. The effect may be explained by the Lorentz force acting on the charge carriers, and the constant can take either sign.

ETTINGSHAUSEN, Albert von,
Austrian physicist, 30 March 1850 Vienna – 9 June 1932 Graz
Lit.: [6b, 22b, 22d, 30, 88]

Hall effect (1879)
The Hall effect is the most important galvanomagnetic effect. If a thin metal plate carries a uniform current and is placed perpendicular to a

magnetic field, the electrons are deflected because of the Lorentz force, and a voltage is produced perpendicular to the field direction and the current direction, which is called the *Hall voltage*. This voltage is directly proportional to the magnetic field, the current flowing, and the geometrical shape of the sample.
The Hall constant is dependent on the conductivity of the material and the carrier mobility in a simple way. It is of the order of a few 10^{-11} m^3 C^{-1}. If the charge carriers are negative, as is assumed here, one speaks of the *normal Hall effect*, but if the carriers are positive, there is an *anomalous Hall effect* (also positive or negative Hall effect). The Hall effect allows one to measure the conductivity and carrier mobility of a material in a simple fashion.
At the same time, the Hall effect provides evidence for the view that free electrons in solids control the conduction mechanism. The → Corbino effect and the → magnetoresistance effect are related to or identical with the Hall effect.
If the plate has good heat insulating properties, the → Ettlingshausen effect (*adiabatic Hall effect*) may be observed simultaneously; if the heat is removed, the isothermal Hall effect may be measured. The Hall effect has been extensively used with modern semiconductors, particularly III-V compounds, in Hall probes, Hall elements, and Hall generators for measurement and control technology.

HALL, Edwin Herbert,
American physicist, 7 November 1855 – 20 November 1938, 1888 – 1921 Professor in Cambridge, Mass.
Hall, E.H., *Am J. Math.* Vol. II
Hall, E.H. *Philos. Mag.* (5) **9, 225**
Lit.: [6b, 22b, 22d, 72 – 78]

Hall effect, anomalous

The Hall effect in ferromagnetic metals is now also called the anomalous Hall effect. It arises from the unsymmetrical scattering of the conduction electrons at the magnetic moments of thin ferromagnetic films. The anomalous Hall constant is dependent on the material type and structure. It can be from 100 to 1000 times larger than the normal Hall constant. The Hall resistance increases linearly with the magnetic field up to the saturation magnetization and then remains constant. At high fields, the resistance decreases slightly on account of the negative normal Hall constant.
Bergmann, G., *Phys. Unserer Zeit* **6,** 177 (1979)

Magnetoresistance effect

The electrical resistance of a magnetic material is altered by a magnetic field (parallel or perpendicular to the current direction). The conduction electrons are more or less scattered at the magnetic spin moments of the atoms, i.e., the electron distribution in the conductor is altered. In a magnetic material this leads to the resistance falling as the field increases, whereas in an unmagnetized semiconductor it leads to an increase in the resistance. Below the Curie temperature the atomic spin ordering increases and the effect becomes larger.
Furthermore, an external magnetic field influences the spontaneous magnetization domains, which also leads to an increase in resistance (magnetoresistance). Hysteresis is observed, and there is a marked increase in the effect at low temperatures.
The effect in semiconductors is closely related to the → Hall effect.
Lit.: [30, 76]

Nernst effect (1887)

In a transverse magnetic field, a longitudinal temperature difference arises when a current flows, which corresponds to the → Peltier effect between magnetized and nonmagnetized materials.

NERNST, Walter Hermann,
German physicist and chemist, 25 June 1864 – 18 November 1941, from 1891 Professor in Göttingen, 1905 – 1933 in Berlin, 1922 – 1924 President of the Physicotechnical State Institute in Berlin
Lit.: [6b, 22d, 30]

Thomson effect (1856)

When a current flows in a transverse magnetic field, there is a change in the electrical resistance or a longitudinal potential difference arises.
THOMSON, Sir William,
British physicist, from 1892 called Lord Kelvin of Largs, 26 June 1824 Belfast – 17 December 1907 Nethergall, Largs, from 1864 Professor in Glasgow
Thomson, W., *Philos. Trans. R. Soc. London* **3,** 737 (1856)
Thomson, W., *Math. Phys. Pap.* **2,** 307 (1884)
Lit.: [30]

Galvanomagnetic effects in semiconductors SC

The experimental relationships are analogous to those for the phenomena described under galvanomagnetic effects (→ conduction effects). An important point is that the coefficients for the individual effects are dependent on the mobilities of both charge carriers (electrons and holes). *Longitudinal* and *transverse effects* may be distinguished.

The following effects occur in a semiconductor wafer when all the components of the magnetic field are active:

The transverse and longitudinal magnetoresistance: this is dependent on the square of the magnetic induction. An anisotropic conductor is necessary.

The transverse planar Hall effect: this is linearly dependent on the magnetic induction. The magnetic field has a component parallel to the current.

Lit.: [53] (additional lit. see Table 48)

Gantmacher effect

→ size effects

Gas ionization

→ nonlinear optical effects

Gay-Lussac experiment (1802) TK

(First Gay-Lussac law)

After preliminary studies by J.A. Charles Gay-Lussac found that in a gas the volume increase per degree Celsius is 1/273 of the volume at 0° C and is constant. The volume of a gas is proportional to the absolute temperature provided the pressure and the number of moles of gas do not alter.

Gay-Lussac experiment (1807) TK

(Second Gay-Lussac law)

In this experiment the volume dependence of the internal energy of a gas is determined. Two vessels connected by a tube that can be closed are completely heat-insulated. One of the vessels is evacuated, and then the gas from the other vessel is allowed to flow into it and the temperature is measured. No temperature change can be detected for an almost ideal gas such as helium. It then follows from the first law of thermodynamics that the internal energy is independent of the volume in an ideal gas.

In a real gas, there is a dependence on the volume, which leads to the → Joule-Thomson effect.

GAY-LUSSAC, Joseph Louis,
French chemist and physicist, 6 December 1778 – 9 May 1850, Professor in Paris

First law

Gay-Lussac, J.L., *Ann. Chim. Phys.* (Paris) **43,** 137 (1802)

Lit.: [50 Vol. III, 1], *textbooks*

Second law

Gay-Lussac, J.L., Mem. d'Arcueil (1807) in Mach, E.: *Principles of Thermodynamics* [in German]. Leipzig 1923

Lit.: [50 Vol. III, 1], *textbooks*

Generalized Hall effects

→ cross effects, → conduction effects

Goos-Hähnchen effect (1943) O

For reasons of continuity and energy, in total reflection a surface wave must arise, which penetrates only a few wavelengths into the optically less dense medium. The finite width of the light beam and diffraction at the boundary lead to a beam shift of a few wavelengths. The wave tests the possibility of propagating in the second medium, so to say. To detect the effect, one side of a prism has a narrow silvered strip. On observing the light reflected from the silver strip and the totally reflected light, a slight beam shift is discovered.

The effect is of significance for couplers and distributors in optical communications and in integrated optics.

→ Schoch effect

GOOS, Hermann Fritz,
German physicist, 11 January 1883 – ?l

Goos, F. and Hähnchen. H. *Ann. Phys.* (Leipzig) (6) **1,** 333 (1947)

Goos, F. and Lindberg-Hähnchen, H., *Ann. Phys.* (Leipzig) (6) **5,** 251 (1949)

Lotsch, H.K.V., *Optik* (Stuttgart) **32,** 116, 189, 299, 553 (1971)

Synder, A.W. and Lore, J.D., *Appl. Opt.* **15,** 1, 236 (1976)

Lee, Y.K. and Wang, S., *IEEE J. Quantum Electron.* **12,** 5, 273 (1976)

Nemoto, S. and Yip, G.L., *IEEE J. Quantum Electron.* **13,** 4, 215 (1977)

Lit.: [6c] (additional lit. see Table 39)

Gorsky effect So

Hydrogen dissolved in a metal migrates from the compressed side to the stretched side when the metal is bent. The concentration gradient then gives rise to an additional nonelastic extension of the metal.

Peisl, H., *Phys. Unserer Zeit* **9**, 39 (1978)
Völkl, J., *Ber. Bunsenges. Phys. Chem.* **76**, 767 (1972)
Lit.: [100, 114]

Gravitation effect

→ light deflection

Gravitational frequency shift R

If an electromagnetic wave is propagating to a gravitational center, there is a *violet shift*, whereas there is a *red shift* if it recedes from the center. The effect is extremely small. In the case of the Sun, one expects a relative frequency shift of 2.12×10^{-6}. The effect is normally concealed by the → Doppler effect. Measurements on the Sun would be expected to show that the violet shift reduces the red shift. This is also observed. The *limb effect* can also be observed: due to the Doppler effect the red shift is only about a third of the gravitational effect when the center of the Sun is observed. The Doppler effect decreases at the limb (edge), and the experimental values tend to Einstein's value for the gravitational frequency shift.

Pound, R. v. and Snider, J.L., *Phys. Rev.* **140**, B 7888 (1965)
Lit.: [6c, 59] (additional lit. see Table 43)

Ground effect F

The flow around a body at ground level produces an uplift different from that in free flow at a large distance from the ground. The vertical flow is impeded near the ground, and the induced resistance is thereby reduced. This leads to an increase in the uplift and a reduction in the resistance.

Major application: air-cushion vehicles (also called ground-effect vehicles).
Lit.: [72, 97, 100]

Gudden-Pohl effect (1920) Lu

In research on electroluminescence, Gudden and Pohl found that the application of an electric field can lead to an increase in light emission. Under certain circumstances, this may continue after optical excitation has ceased.
→ Déchêne effect

GUDDEN, Bernhard Friedrich,
German physicist, 14 March 1892 – 3 August 1945, Professor in Prague

POHL, Robert Wichard,
German physicist, 10 August 1884 – 5 June 1976, Professor in Göttingen
Lit.: [6d], dictionaries

Guest-host effect

→ liquid crystal effects

Gunn effect (1963) SCC

High-frequency current oscillations can occur in a homogeneous semiconductor at field strengths above 2 kV cm^{-1} if the conduction band has several energy minima in which the electrons have very different mobilities. The name Gunn effect is given to negative conductivity as a *bulk effect* in such semiconductors.

The explanation for the effect lies in the band structure of the semiconductor. The conduction band of GaAs, for example, has not only the main minimum but secondary minima, and the energy separation between the main and secondary minima is 0.36 V. The electrons collect in the main minimum at room temperature in the absence of a field. Because of their high mobility in this region of the conduction band, the electrons very readily acquire energy from an applied electric field, and consequently an appreciable fraction of the electrons will pass directly or via the → tunnel effect to the secondary minimum. The lattice is involved in this process in maintaining the laws of conservation of energy and momentum. The mobility in the secondary minimum is less by about a factor 20, so the conductivity decreases sharply, and there is an abrupt reduction in the current through the semiconductor.

This negative conductivity or resistance is the basis for oscillation and amplification. A more detailed examination shows that space-charge instabilities occur as a consequence of the negative differential electron mobility, which is related to the negative conductivity. These electron clouds (domains) move in an electric field with a speed of about 10^7 cm s^{-1}. The time separation between the acicular current pulses or the oscillation fre-

quency is controlled by the electrode separation and the velocity: for a separation of 20 μm, the frequency is about 10 GHz for GaAs.
Components that utilize this effect are called Gunn diodes (electron-transfer components) and have been produced as small devices of medium power for the microwave range.
→ avalanche effect, → Zener effect

GUNN, John Battiscombe,
British physicist, 13 May 1928
Wunn, J.B., *Solid State Commun* **1,** 88 (1963)
Heywang, W., *Phys. Unserer Zeit* **2,** 130 (1971)
Lit.: [52, 53, 74, 75, 92, 116]

Gurevich effect

→ thermoelectric effects, → phonon-drag effect

Gush effect

→ Lenard effect

Gyromagnetic effects EM

The gyromagnetic effects link the mechanical angular momentum of an atom or a body with the magnetic moment, arising mainly due to electron spin (angular momentum). Any change in the total mechanical angular momentum of the electrons must correspond to a change in the angular momentum of the body. Therefore, the ratio of the magnetic moment to the mechanical angular momentum can be determined by measuring gyromagnetic effects. The magnetic moment associated with an electron is larger by a factor two than that which would be assigned to an orbital motion with the same angular momentum. Experiments thus allow conclusions to be drawn on the contributions from the orbital and spin moments to the magnetization of a material.
In general, the gyromagnetic ratio for a material may be distinguished from the gyromagnetic ratio for the orbital angular momentum via the Landé factor g, which can be somewhat larger but also much smaller than two. Calculations on the value of g for electrons and muons made from quantum electrodynamics have been confirmed by experiment.
→ Barnett effect, → Einstein-de Haas effect
Lit.: [6b, 16, 50 Vol. IV, 4, 69, 6d, 7] (additional lit. see Table 26)

Haas effect (1951) Me

If sound waves are produced by two loudspeakers in such a way that the sound derived from the second loudspeaker is delayed in time relative to that from the first, the echo source with delay times between one and 30 milliseconds cannot be detected if the intensity is less than 10 dB more than that of the primary sound.

HAAS, Arthur Erich,
Austrian physicist, 30 April 1884 – 20 February 1941, Professor in Leipzig, Vienna, and from 1936 in the USA
Haas, H., *Acustica* **1,** 49 (1951)
Lit.: [76] (additional lit. see Table 35)

De Haas-van Alphen effect (1930) SS

This effect serves to define the *Fermi surface* in a solid. The Fermi surface is defined by considering all the electrons in momentum space whose energy is equal to the Fermi energy. This is defined as the limiting energy at absolute zero, at which all the electron states in a metal or degenerate semiconductor are occupied up to the limiting Fermi energy, while beyond that all states are empty. For free electrons a sphere is obtained, whereas the Fermi surface in a metal is extremely complicated.
Below 4 K, a strong magnetic field of some tesla is applied to a single-crystal solid. This deflects the electrons moving on the Fermi sphere to the Equator, where they vanish. The resulting fluctuations in the magnetic susceptibility as a function of the field produce an alternating voltage, whose frequency is measured, and from which the Fermi surface itself can be deduced.
Many properties of metals and semiconductors can be derived quantitatively from a knowledge of the Fermi surface. A knowledge of the surface is therefore of considerable significance. Various methods are used to define the Fermi surface:
→ cone effect, → magnetoacoustic effect, → anomalous skin effect, → Azbel-Kaner effect, magnetoresistance, → Shubnikow-de Haas effect, → size effect, → Sondheimer effect

DE HAAS, Wander Johannes,
Dutch physicist, 2 March 1878 – 26 April 1960
de Haas, W.J. and van Alphen, P.M., *Proc. Amsterdam Acad.* **33,** 1106 (1930); *Leiden Comm.* **212 (1931)**
Lit.: [1, 3, 27, 29, 43, 71] (additional lit. see Table 51)

Hafele-Keating experiment
→ time effects

Hall effect
→ galvanomagnetic effects

Hanle effect (1925) Mo
A Hanle effect is the influence of a magnetic field on the polarization in light emission (1924). If, for example, mercury atoms are excited with linearly polarized light, linear polarization is observed perpendicular to the direction of incidence and to the electric vector of the excitation. A magnetic field applied in the observation direction produces a rotation of the plane of polarization and depolarization which increase with the field strength.
An atom excited by linearly polarized light can be described classically as a linear oscillator, which performs a precession movement in the magnetic field, decaying in accordance with its lifetime. If the degree of polarization is plotted as a function of the magnetic field (the Hanle curve), the product g_τ can be determined directly from the half-width, and from this one can derive one of the two quantities, namely, lifetime τ or Lande factor g, for the excited state.
The effect can be interpreted quantum mechanically as the lifting of the degeneracy of coherently excited states. On account of the natural width of the excited level (of the order of $1/\tau$), the Zeeman splitting levels overlap in the region of the crossing point in a magnetic field of zero (Null field level crossing). Within this degeneracy, the states emit with a strict phase relationship, which leads to an interference term and thus to a change in radiation polarization. This coherence derives from states in individual atoms, so there is no influence from the Doppler effect. This was the *first* spectroscopic method free of the Doppler effect. Level crossing occurs in a static magnetic field, in a static electric field, and in the dynamic alternating electric field of a high-intensity somewhat frequency-shifted laser (optical Hanle effect).
Also at higher field strengths, atomic coherences arise from the crossing of the magnetic sublevels of the fine-structure or hyperfine-structure states. The Hanle effect is therefore a special case of general level crossing.
It has numerous applications, such as the measurement of lifetimes or g factors, and the determination of the degree, transfer, and relaxation of alignment, as well as the determination of orientation in collisions and extraterrestrial magnetic fields.
→ nuclear Hanle effect

HANLE, Wilhelm,
German physicist, 13 January 1901, Professor in Giessen

Hanle, W., *Z. Phys.* **30**, 93 (1924)
Hanle W., *Z. Phys.* **35**, 346 (1926)
Kaftandjian, V.P. and Klein, L., *Phys. Lett.* **62A**, 317 (1977)
Hanle, W., Kaftandjian, V.P., and Klein, L., *Phys. Lett.* **65A**, 188 (1978)
Hanle, W. and Pepperl, R., *Phys. Bl.* **27**, 19 (1971)
Kastler, A.: *Phys. Bl.* **30**, 394 (1974)
Corney, A.: *Atomic and Laser Spectroscopy*. Clarendon Press, Oxford 1977

Heal effect
In a modern X-ray tube, the electron beam and the anode form an angle. Since the X-rays are produced not only at the surface but also at a finite depth in the anode, the proportion of radiation that emerges at glancing angles to the anode is reduced by the anode material itself. This is called the heal effect. Thus, the energy distribution in the beam varies, depending on the angle of emergence.
To reduce the effect, which interferes with clinical applications, the anode angle and film distance are increased, and large objects are placed on the cathode side.
Lit.: [36], dictionaries

Heilmeier memory effect
→ liquid-crystal effects

Hertz effect (1886) EM
A linearly polarized electric wave is attenuated to a certain extent on passing through a grating made of metal rods, the effect being dependent on the orientation of the electric vector to the rod direction. A minimum occurs when the electric vector is perpendicular to the rod direction (the Hertz effect). The incident wave excites the rods to radiate electric waves, and these interfere with the incident waves. The Hertz effect must be al-

lowed for in communication at short and ultrashort wavelengths over land, for example, in communication over forested areas.

HERTZ, Heinrich Rudolph,
German physicist, 22 February 1857 – 1 January 1894, from 1885 Professor in Karlsruhe, after 1889 in Bonn

Hertz's experiments (1888 – 1894) EM

In these experiments, Hertz demonstrated the existence of electromagnetic waves and their similarity or identity to light. In particular, the following experiments are involved: *reflection, refraction, diffraction,* and *polarization of electromagnetic waves.*

Hertz, H., *Ann. Phys.* **31**, 421 (1887)
Hertz, H., *Ann. Phys.* **34**, 155, 551, 609 (1888)
Hertz, H., *Sitzungsber. K. Preuss Akad. Wiss.*, Berlin 1888, vol. II of Collected Papers, Leipzig (1894)
Lit.: [100, 103, 109, 114]

High-field effects

→ avalanche effect, → Gunn effect, → Zener effect

Holography (1947/1948) L

Holography goes back to Dennis Gabor (1947). He demonstrated picture construction with coherent light without the use of focussing optics. In simple terms, a hologram is an interference pattern which is produced from a wave field arising from an object by the use of a reference wave, and is stored on a film.

By reversing the wave path used in the recording, a virtual picture is obtained that can be viewed from different angles and thus has a three-dimensional appearance. On readout by irradiating the hologram, the object wave is completely reconstructed. One therefore sees a three-dimensional picture of the object. Hence, a hologram contains not only amplitude information (photography) but phase information.

In *real-time holography*, the information stored in the hologram about the object is compared with that for the object under load; the hologram is adjusted to its original position, and then the original and the virtual picture fully coincide. If the object is loaded (mechanically, thermally, or electrically), deviations occur from the interference pattern stored in the hologram. The resulting interference-fringe pattern on the object can be interpreted.

Double-illumination holography relies on the fact that several wavefronts can be stored in a hologram. A holographic recording is made of the object in the unloaded state, but this is interrupted after half the exposure time, and then the object is loaded and the exposure is completed. A picture of the object with an interference pattern corresponding to the loading is obtained and evaluated.

The following applications of holographic interferometry are important: nondestructive material testing (quality testing), shape comparison, vibrational analysis, and the examination of phase objects.

The size of the object cannot be chosen arbitrarily, since limits are set to the reasonable use of interferometric holography by the coherence length of the light, the optics, and lastly the experimental effort involved for large objects. On the other hand, middle-sized objects can be examined, such as vehicle tyres, parts of vehicle bodies, motors, etc.

GABOR, Dennis,
Hungarian-British physicist, 5 June 1900 – 9 February 1979, from 1958 Professor in London

Gabor, D., *Nature* (London) **161,** 777 (1948)
Gabor, D., *Proc. R. Soc. London Ser.* A **197,** 454 (1949)
Lit.: [26, 33, 57] (additional lit. see Table 31)

Hopkinson effect (1890) EM

At low field strengths the permeability of a ferromagnetic material measured as a function of temperature attains a maximum a little below the Curie temperature, which is called the Hopkins effect. No maximum is found in strong fields. The explanation of the effect is that the crystal anisotropy and the spontaneous magnetization have different gradients near zero. The crystal anisotropy is very small at lower temperatures, whereas the spontaneous magnetization decreases sharply only in the neighborhood of the Curie temperature.

HOPKINSON, John,
British electrical engineer, 27 July 1849 Manchester – 27 August 1898 Switzerland

Hopkinson, J., *Proc. R. Soc. London, Ser. A* **48,** 1 (1890)
Lit.: [26] (additional lit. see Table 26)

Hopping effect SC

In an amorphous solid, the lack of long-range order means that the pronounced conduction and valency bands of the crystalline state are replaced by more or less localized states, which overlap in energy. There are therefore no energy gaps. The electrons move in the electric field between these localized states. The motion is possible because of the → tunnel effect or hopping over potential barriers. This conduction mechanism is prominent in amorphous semiconductors.

Lit.: [91] (additional lit. see Table 48)

Hubble effect (1929) As

The red shift of spectral lines from extragalactic star systems is proportional to the distance between the systems to a first approximation (Hubble effect). As the red shift is interpreted as → optical Doppler effect, the velocity of recession of the star system may be derived; it increases linearly with distance. The coefficient of proportionality (the Hubble constant) has a value of about 100 km s^{-1} per million parsec.

HUBBLE, Edwin Powell,
American astronomer, 20 November 1889 – 28 September 1953, from 1919 at Mt. Wilson Observatory in Pasadena

Lit.: [6c] (additional lit. see Table 22)

Hughes effect (1935) EM

Asymmetry in the alternating-current hysteresis occurring in thin sheet cores free from air gaps made of permalloy or Mu metal is called the Hughes effect. The effect is prominent when the maximum field strength is in the region of the maximum direct-current permeability, and it is ascribed to polarization in the laminations. Interference due to this effect can be eliminated by using a gradually decreasing strong magnetic field strength.

HUGHES, Edward Charles,
American physicist, 13 February 1901

Hughes, E., *J. Inst. Electr. Eng.* **79**, 213 (1936)

Lit.: [26] (additional lit. see Table 26)

Huygens' principle

→ diffraction

Hyper-Raman effect (1965) L

If a system is exposed to the beam from a giant-pulse laser (wave number ν_0) whose intensity is above the threshold for nonlinear interaction, radiation is observed not only with frequency $2\nu_0$ but with frequency $2\nu_0 \pm \nu_M$, where ν_M is the frequency of a transition in the scattering molecule. The scattering at $2\ \nu_0$ is called the hyper-Raman effect.

Long, D.A.: *Raman Spectroscopy*. McGraw-Hill, New York 1977

Brunner, W., and Junge, K.: *Laser Engineering: an Introduction* [in German]. Hüthig, Heidelberg 1982

Lit.: [98, 100, 106]

Hysteresis effect

→ liquid-crystal effects, → relaxation effect

I effect

→ substituent effects

Immersion effect (1921) AQ I

The light-emitting electrons in an atom can penetrate deep into the remainder of the atom. There they experience a strong binding potential, which reduces the energy and increases the binding energy. The effect is larger the bigger the core or the heavier the atom, and also the smaller the electron orbital angular momentum. The remaining electrons in the shell are subjected to the full nuclear field and so the immersion effect becomes less important.

Lit.: [90] (additional lit. see Table 20)

Induced scattering

→ nonlinear optical effects

Induction effect

→ dipole effects, → substituent effects

Interference (1802) G

This is the name given to the general property observed in 1802 by Thomas Young of monochromatic coherent waves to be strengthened or attenuated on superposition in accordance with their phase relationship; characteristic intensity maxima and minima result. Depending on the number of interfering waves involved, a distinc-

tion is drawn between double-beam and multiple beam interference. There are numerous applications of interference effects in measurement technology. Since the discovery of the → laser, interferometric measurement methods have gained in significance. Interference phenomena are also found with elementary particles because of wave-particle dualism.
→ Fizeau's experiment, → Michelson's experiment, → Sagnac effect
Lit.: *textbooks, especially* [6c, 9, 22c, 33, 38, 61, 93, 98, 106]

Interionic interaction
→ dispersion effects

Internal absorption
→ Auger effect

Internal photoelectric effect
→ photoeffects

Internal photoionization
→ Auger effect

Inverse Compton effect
→ Compton effect

Inverse magnetostriction
→ magnetic contraction effect

Inverse Peltier effect
→ thermoelectric effects

Inverse thermal-diffusion effect
→ thermal diffusion

Inverse Thomson effect
→ thermoelectric effects

Inverse Zeeman effect
→ Zeeman effect

Ioffe effect (1924) So
If an ionic crystal, such as rocksalt, is exposed simultaneously to a mechanical stress and a solvent, this can lead to an increase in the plasticity and strength of up to 25 times the dry strength.

IOFFE, Abraham Federovich,
Russian physicist, 29 October 1880 – 14 October 1960, from 1918 Professor in Leningrad
Ioffe, A.F., *Z. Phys.* **22**, 17 (1924)
Lit.: [72] (additional lit. see Table 50)

Isothermal galvanomagnetic effects
→ galvanomagnetic effects

Isothermal Hall effect
→ galvanomagnetic effects

Isotope effect (1950) S
In isotopic superconductors it has been observed that the transition temperature and the critical magnetic field are dependent on the mass of the isotope. The effect demonstrates the significance of the interaction between electrons and lattice vibrations in the superconducting state.
Lit.: [11] (additional lit. see Table 44)

Isotope effect TK
In the diffusion and evaporation of isotopic mixtures, one finds that the light isotopes diffuse and evaporate more rapidly. A consequence is that chemical reaction rates for different isotopes vary. The effect is used in separating isotopes.
Lit.: [76]

Jaccarino-Peter effect (1962, 1984)
(compensation effect) S
In a ferromagnetic metal, the spins of the conduction electrons are polarised owing to an exchange interaction with the magnetic ions and therefore no superconductive state – formed of Cooper pairs – can exist. If the exchange interaction in a ferromagnet, however, has a weak anti-ferromagnetic character (i.e., exchange integral $J < 0$), the polarisation of the conduction electrons can be compensated by an external magnetic field of appropriate strength, and superconductivity can be induced. This can occur, however, only in materials in which the diamagnetic action of the compensating field does not lead to the collapse of the superconductivity, i.e., in materials with extremely high critical fields B_{c2} – a property that makes these materials interesting for technological applications. This compensation effect, predicted for weak ferromagnets by V. Jaccarino and M. Peter in 1962, can also occur in paramagnets with negative exchange interaction – this case dectected experimentally for the first time in 1984 by Meul et al. The B_{c2}-temperature phase diagram comprises two separate superconductive regions, the

usual one at small magnetic fields and a second in the high field region, induced by the compensation effect, with anomalous behavior of the flow-line grid (anti-vortices).

Jaccarino, V., and Peter, M., *Phys. Rev. Lett.* **9**, 290 (1962)
Maul, H.W., et al., *Phys. Rev. Lett.* **53**, 497 (1984)
Fischer, O., et al., *Phys. Rev. Lett.* **55**, 2972 (1985)

Jahn-Teller effect (1937) SS

In a crystal defect the degeneracy of the electron or hole states is lifted because of the deformation of the surrounding lattice. A distinction is drawn between the *static* and *dynamic Jahn-Teller effects*. In cases where a strong interaction between the lattice and the defect exists the equilibrium positions of the lattice atoms alter in such a way that the symmetry is reduced and therefore the degeneracy is lifted. The energy of the complete system is lower than that without the defect (static case). If the interaction between the lattice and the defect site is weaker, the degeneracy is lifted due to the interaction of the defect with the phonons (dynamic case). In that case, the zero-point energy of the lattice vibrations is comparable with the energy change caused by the lattice deformation.
The Jahn-Teller effect is of basic significance for the chemical bond. It gives information on which structures and which electron states have the lowest energy.

TELLER, Edward,
Hungarian-American physicist, 15 January 1908 Budapest
Sturge, M.D., *Solid State Phys.* **20**, 91 (1967)
Lit.: [27, 74, 75] (additional lit. see Table 51)

Johnsen effect

→ shot effect

Johnsen-Rahbeck effect SC

Whun a current flows between a metal and a semiconductor, the electrostatic attraction between the electrodes acting as capacitor plates produces a force. This forces arises because the metal and the semiconductor are in contact only at a few points, through which the current flows when a voltage is applied. The result is that many small capacitors are created, whose forces add up when the separations are small. In the case of a system consisting of a metal and a semiconducting liquid, the effect is known as the *Winslow effect*. This effect can be used, for example, in loudspeakers: a rotating semiconductor is partly surrounded by a metal strip. A voltage between the roller and the metal strip increases the friction and thus the tension exerted on the membrane.

JOHNSEN, John Bertrand,
Swedish-American physicist, 2 October 1887 – ?
Lit.: [6b, 69, 2nd Ed., 50 Vol. 4, part 1] (additional lit. see Table 48)

Josephson effects (1962) S

In 1962, Josephson predicted that, by analogy with the normal electron-tunnelling effect, a tunnel effect should occur for Cooper pairs in two superconducting materials separated by a thin insulating layer (about 1 mm). The magnetic field in the insulating layer should be only a few A cm^{-1}. Electrons with antiparallel spins are coupled as quasiparticles, called Cooper pairs.
The tunnelling current of Cooper pairs is lossfree provided it does not exceed a critical value. The tunnelling junction then behaves as a superconductor. The tunnelling current oscillates as a function of the applied magnetic field with the number of flux quanta involved (the *DC Josephson effect*). A field of about 1 A cm^{-1} serves to switch the tunnelling current.
If an additional voltage is applied to the junction that is less than that required to overcome the energy gap, an alternating current is produced which is dependent only on the fundamental constants h and e (the *AC Josephson effect*). The Josephson effect thus allows very exact measurement of Planck's constant h. At present, attempts are being made to use the features of the Josephson effect in fast superconducting memories.

JOSEPHSON, Brian David,
British physicist, 4 January 1940 Cardiff (Wales), from 1972 Professor in Cambridge
Josephson, B.D., *Phys. Lett.* **11**, 251 (1962)
Josephson, B.D., *Phys. Rev. Lett.* **1**, 251 (1962)
Fritsch, G., *Phys. Unserer Zeit* **6**, 126 (1975)
Shapiro, S., *Phys. Rev. Lett.* **11**, 80 (1963)
Lit.: [6d, 11, 15, 22d, 27, 44, 74, 75, 91] (additional lit. see Table 44)

Joule effect (1841) EM

The heating of a conductor when current flows, is called the Joule effect. The heat is proportional to the resistance of a conductor, the square of the current, and time.
The Joule effect masks the Benedicks effects and the Thomson effects.
→ thermoelectric effects

JOULE, James Prescott,
British physicist, 24 December 1818 Salford, Lancashire – 11 October 1889 Sale, Cheshire
Joule, J.P., *Philos. Mag. Ser. III* **19**, 260 (1841)
Lit.: [28b] (additional lit. see Table 26)

Joule effect (1842)

→ magnetostriction

Joule-Thomson effect (1847/1952) TK

(the throttle effect)
In a real gas adiabatic expansion at a throttle produces a temperature difference between the flows before and after the throttle. The pressure is high and the volume is small before the throttle, while after it the pressure is small and the volume is large. Joule and Thomson found the cooling after the throttle to be proportional to the pressure difference and inversely proportional to the square of the temperature.
In certain cases, there is an increase in temperature on throttle expansion, which is called the *inverse Joule-Thomson* or *throttle effect*. The theory of the Joule-Thomson effect also leads to two further expressions: the *differential* and *integral Joule-Thomson effects*. By the first, one means the temperature change for a differential pressure change at constant heat content, while the temperature difference corresponding to a finite pressure difference (the throttle cooling) is called the integral Joule-Thomson effect. The effect is the basis for large-scale liquefaction of gases (the *Linde method*).
The names Joule-Thomson effect and throttle effect are generally used synonymously. By isothermal throttle effect, one means the change in enthalpy with pressure at constant temperature, which is the amount of heat one has to supply to the throttle in order to eliminate the Joule-Thomson effect, i.e., to keep the temperature constant.
→ Gay-Lussac experiment (1807)

THOMSON, William Lord Kelvin,
1824 – 1907,
Joule, J.P. and Thomson, W., *Philos. Mag.* **4**, 481 (1952)
v. Linde, G., *D.R.P.* **88**, 824 (1895/1896)
v. Linde, G., *Ber. Dtsch. Keram. Ges.* **32**, 925 (1899)
Lit.: [6a, 22a, 23, 50 Vol. III, 1] [70 Vol. I] (additional lit. see Table 54)

Kelvin-Helmholtz effect Sy

(Kelvin instability)
If a change in velocity occurs in a fluid flowing between two layers, instabilities arise at the boundary layer. This effect is also observed in flowing plasma. A magnetic field parallel to the flow direction stabilizes these instabilities if the relative velocity does not exceed a certain value.
Chandrasekhar, S.: *Hydrodynamic and Hydromagnetic Stability*. Oxford University Press, Oxford 1961
Cap, F.: *Introduction to Plasma Physics* [in German], Vol. I, II, III. Vieweg, Brunswick 1970, 1972

Kerr effect (1875) Eo

(electrooptic birefringence)
In this effect, an optically isotropic material becomes optically anisotropic in a homogeneous electric field. The refractive index differences are proportional to the square of the field strength. One usually employs a liquid such as benzene, carbon disulfide, or nitrobenzene. A distinction is drawn between positively and negatively birefringent substances, according to whether the ordinary or extraordinary ray has the higher speed. If the molecules are aligned by the field, we have an *orientation Kerr effect*. If the electrons in the atoms or molecules are affected, we have an *electronic Kerr effect*.
The effect is largest in liquids. In solids it is smaller by at least an order of magnitude, and in gases by three orders of magnitude. The Kerr effect is used in switching and controlling light, e.g., in high-speed photography or in giant-pulse lasers. Higher modulation frequencies, however, are provided primarily by crystals.

KERR, John,
British physicist, 17 December 1824 – 18 August 1907, mathematics teacher in Glasgow
Kerr, J., *Philos. Mag.* (4) **50**, 337, 446 (1875)
Lit.: [6c, 9, 22c, 19, 26, 57, 69, 71, 77, 50 Vol. II, 22] (additional lit. see Table 27)

Kikuchi effect (1928) SS

Electron diffraction at fairly thick crystal layers produces not only the usual interferences but dark and light almost straight lines. Generally, a light line and a dark line are parallel to each other. They arise through the interference and diffraction of electrons moving diffusely in the crystal, where they undergo inelastic scattering and diffraction at lattice planes. Each lattice plane produces a cone of diffracted radiation, which is related to an absorption cone. If the absorption of the diffuse primary beam and of the interference beams were the same, the effect would vanish.
→ Kossel effect

KIKUCHI, Seishin,
Japanese physicist, born 1902
Kikuchi, S., *Jpn. J. Phys.* **5,** 83 (1928)
Lit.: [29a, 37] (additional lit. see Table 51)

Kinetic effects

→ transport effects

Kirk effect (1962) SCC

The charge carrier transit time through the collector space-charge zone is controlled by the width of the latter and by the drift speed.
At elevated frequencies, the effective base shifts into the collector zone, since the injected minority carrier density becomes of the order of the base doping. This reduces the limiting frequency. The effect is used to adjust transistor gain and collector parameters.
Kirk, C.T., *IEEE Trans. Electr. Rev.* EO-9, 164 (1962)
Lit.: [84 Vol. III] (additional lit. see Table 49)

Kirkendall effect 1942 SS

The diffusion of components and the related shift of interfaces in alloys is called the Kirkendall effect. The alloy components diffuse independently between lattice sites and vacancies. The shift is of the order of 10^{-1} to 10^{-2} nm.
The Kirkendall effect shows that diffusion is possible in solids only by the motion of atomic defects.
Kirkendall, *Trans. Am. Inst. Min. Metall. Engrs.* **147,** 104 (1942)
Lit.: [36, 87 Vol. VII. 1] (additional lit. see Table 51]

Kirlian effect (1939) EM

(Lichtenberg figures 1777)
This effect relates to discharge figures, for example, of sheets or other objects, produced by high static voltages. The effect can be used, for example, to photograph an object emitting or reflecting in the infrared region with photosensitive semiconductors.
→ corona effect

LICHTENBERG, Georg Christoph,
German author and physicist, 1 July 1742 – 24 February 1799, from 1769 Professor in Göttingen
Pehek, J.O., Kyler, H.J., and Faust, D.L., *Science* **194,** 263 (1976)
Pehek, J.O., Kyler, H.J., and Faust, D.L., *Phys. Unserer Zeit* **8,** 66 (1977)
Lit.: [27, 50, Vol. IV, part 1] (additional lit. see Table 26)

von-Klitzing effect (1980) Ga

(quantum Hall effect)
When the Hall voltage is measured at low temperatures (liquid helium) and high magnetic fields (about 15 T) on a semiconductor inversion film or on a thin semiconductor film, there is quantization of the Hall voltage or the Hall resistance. The reason lies in the almost complete quantization of the two-dimensional electron gas related to the surface field or the thin film. It can be shown that the measured values of the Hall voltage are dependent only on the fine-structure constant α and the speed of light c. The geometrical shape of the sample is not involved.
To obtain the two-dimensional electron gas, high-quality MOS field-effect transistors with (100) orientation are used. Additional electrodes fitted to the channel in this transistor are used to measure the Hall voltage.
At low temperatures the transverse magnetic field gives the Landau energy levels that correspond to circles in the two-dimensional case (→ Azbel-Kaner effect). The electrons can move only along such paths. In principle, these are → Shubnikov-de Haas oscillations, in which the → magnetoresistance oscillates as a function of the magnetic field. The Hall constant contains the number of free charge carriers per unit volume. This applies equally for three-dimensional and two-dimensional samples. In the two-dimensional case, the Landau quantization of the en-

ergy levels means that the Hall conductivity is quantized in units of e^2/h.
Von Klitzing's measurements gave a value for α with an uncertainty of 10^{-6}. An important point is that this value is independent of quantum field theory. Further, it provides a means of measuring the resistance unit, the Ohm, by means of the quantized Hall resistance.
Klitzing, K., Dorda, G., and Pepper, M., *Phys. Rev. Lett.* **45**, 494 (1980)
Landwehr, G., *Phys. Bl.* **37**, 59 (1981) (additional lit. see Table 30)

Knight shift (1949) AQ II

The nuclear resonance frequency in a metal is shifted to higher values relative to that of the same nucleus in a diamagnetic compound.
This effect is due to the paramagnetism in the conduction electrons, which produces additional internal fields at the nucleus via the Fermi contact contribution to hyperfine interaction. The shift is virtually temperature-independent (as is the paramagnetism in the conduction electrons) and increases (as does the Fermi contact interaction) with the probability density of the electrons at the nucleus, and thus with the nuclear charge Z. The Knight shift therefore enables conclusions to be drawn on the electron distribution in a metal. In addition, it can be used to investigate the superconducting state, since the paramagnetism of the conduction electrons tends to zero as Cooper pairs become more predominant, and so the Knight shift vanishes.
Townes, C.H., Herring, C., and Knight, W.D., *Phys. Rev.* **77**, 852 (1950)
Knight, W.D., in Seitz and Turnbull (eds.): *Solid State Physics,* Vol. 2. Academic Press, New York 1956

Knudsen effect (1910) TK

The term *effusion* describes the flow of gases through small pores or capillaries whose diameter is small by comparison with the mean free path. Two vessels are linked by a porous plate or a capillary. If one side of the plate or capillary is heated, a pressure difference is found between the two vessels.
Knudsen distinguished between two cases: if the entire vessel up to the openings of the capillaries is heated, gas is lost, and the density in the vessel is reduced, because the pressure increases less than the temperature.
If only the capillary opening is heated, gas enters the vessel, because the pressure must increase, corresponding to the temperature of the opening, without the temperature in the vessel rising.
If the ends of the capillary are connected to wide glass tubes, a gas flow can be produced by heating one capillary end.
Dufour discovered the inverse effect: on diffusion through a porous plate (flow through a capillary), there is a temperature rise at the entrance.
KNUDSEN, Martin Christian,
Danish physicist, 15 February 1871 – 27 May 1949 Fyn
Knudsen, M., *Ann. Phys.* **31**, 205, 633 (1910)
Dufour, L., *Arch. Sci. Phys. Nat.* (Geneva) **45**, 9 (1872)
Dufour, L., *Pogg. Ann.* **148**, 490 (1873)
Lit.: [50 Vol. III, 2] (additional lit. see Table 54)

Kohn effect SS

When the wave humber vector of phonons travelling through a material corresponds to the diameter of the Fermi sphere, the phonons tend to produce standing waves, as in Bragg reflection at lattice planes. This produces a sharp change in the phonon dispersion curve and is called the Kohn effect. The effect is normally very small; it becomes somewhat larger if the Fermi surface deviates considerably from a spherical shape. The Kohn effect can be used in evaluating the Fermi surface, as can the de Haas-van Alphen effect, but it does not have the same significance as the latter on account of its small magnitude. It has been observed in lead and aluminium.
KOHN, Walter,
Austrian-American physicist, 9 March 1923 Vienna
Lit.: [21] (additional lit. see Table 51)

Kondo effect (1964) SS

In certain metals containing small amounts of magnetic impurities, for example, about 10^{-3} % iron in very pure copper, a resistance minimum instead of a constant or decreasing resistance is found at low temperatures between 10 K and 40 K. After the minimum the resistance then rises as the temperature decreases. This effect is due to interaction between the conduction-electron spins and the spins of the foreign atoms.
KONDO, Yoshiva, 1924
Kondo, J., *Prog. Theor. Phys.* **32**, 1589 (1964)
Lit.: [1, 6d, 20, 75] (additional lit. see Table 51)

Kossel effect (1934/1935) SS

In the diffraction of a spherical X-ray wave at a crystal, which is simultaneously the anticathode in an X-ray tube, an emission occurs in which the lattice atoms act as independent sources of monochromatic radiation.
The diffracted and emitted rays form a system of double cones with respect to the lattice planes, and the intersections with a photographic film enable the crystal lattice constants to be determined.
The effect is related to the → Kikuchi effect.

KOSSEL, Walther,
German physicist, 4 January 1888 – 22 May 1956, Professor in Kiel, Danzig, and Tübingen
Lit.: [72] (additional lit. see Table 51)

Kundt effect (1884) Mo

An extraordinarily large magnetic rotation of the plane of polarization occurs when polarized light passes through very thin films of ferromagnetic materials. Very pure films of Fe, Ni, and MnBi give various values for the Kundt constant of the order of 7 x 10^5 to 8 x 10^5 deg cm.
(→ magnetooptic effects)

KUNDT, August Adolph Eduard Eberhard,
German physicist, 18 November 1839 – 21 May 1894, Professor in Zürich, Würzburg, Strasbourg, and Berlin
Kundt, A., *Wiedem. Ann.* **23,** 228 (1884)
Kundt, A., *Wiedem. Ann.* **27,** 191 (1885)
Lit.: [3, 22c, 89] (additional lit. see Table 37)

Lamb shift (1947) AQ II

(Bethe effect)
The Lamb shift is a very small displacement of the spectral lines of the hydrogen atom. According to Dirac's theory, the $2^2S_{1/2}$ and the $2^2P_{1/2}$ levels should coincide. Using high-frequency spectroscopy, Lamb and Retherford were able to determine that the singlet term lies slightly above the triplet one. This is explained from the interaction of the electron with its own virtual radiation field, particularly with the vacuum polarization of the electromagnetic field, which produces a fluctuation or wobbling movement in the electron.
The magnitude of the Lamb shift provides striking conformation of the theoretical predictions from quantum electrodynamics.

LAMB, Willis Eugene,
American physicist, 12 July 1913 Los Angeles, Professor at Yale University
BETHE, Hans Albrecht,
German-American physicist, 2 July 1906, from 1935 Professor in Ithaca (USA)
Lamb, W.E. and Retherford, Robert G., *Phys. Rev.* **72,** 241 (1947)
Bethe, H.A., *Phys. Rev.* **72,** 339 (1947)
Schwinger, J. (ed.): *Quantum Electrodynamics.* Dover Publ. Ltd. 1958
Lit.: [6d, 38 Vol. IV, 74, 75] (additional lit. see Table 21)

Langmuir effect (1923) AQ II

(thermionic effect)
In the Langmuir effect, neutral atoms under vacuum are ionized at hot metal surfaces. The electron work function for the metal must be greater than the atomic ionization energy.
The effect is used in producing high-intensity positively and negatively charged ion beams, as well as in neutral-atom detectors (*Langmuir-Taylor detector*).

LANGMUIR, Irving,
American physical chemist, 31 January 1881 – 16 August 1957, from 1909 industrial physicist and chemist at General Electric Co., Schenectady
Langmuir, I., *Science* **57** (1923)
Lit.: [75] (additional lit. see Table 21)

Lasers (1960) L

Laser is an abbreviation for **l**ight **a**mplification by **s**timulated **e**mission of **r**adiation. Lasers therefore are amplifiers and particularly oscillators in the optical range. The difference to filament lamps or discharge lamps, which use spontaneous emission (in general, from thermal sources), is that the laser employs *induced emission* and nonthermal sources. The resulting radiation is monochromatic and has time and spatial coherence. As in high-frequency engineering, amplification requires an amplifying element and feedback. In the microwave and optical ranges, such elements are provided by invertible materials and optical cavities.
Stimulated emission is basic to the operation of lasers (and also of masers, which is the acronym for **m**icrowave **a**mplification by **s**timulation **e**mission of **r**adiation).

Light emission is the converse of absorption. When light passes through an optical material, it is exponentially attenuated or absorbed. This means that some of the quanta give up their energy to the atoms, which are excited; electrons are raised from the ground level to higher ones. The light leaving the material has an intensity lower than that on input. The stimulated atoms give up their energy spontaneously and in a statistically random fashion in the form of light or heat. The spontaneous emission (electron transition from a higher level to a lower one) is not dependent on an external radiation field.

In 1916, Albert Einstein showed that in addition to spontaneous emission there must be a further process, called induced or stimulated emission. If the excited atoms or molecules give up their energy as light in response to an external radiation field, one speaks of induced emission. Absorption, induced emission, and spontaneous emission are in equilibrium under normal conditions. For atoms or molecules this means that the energy levels are populated according to a Boltzman distribution: the ground state is fully occupied by electrons, while the populations of the higher energy levels decrease exponentially.

If the system is supplied with energy, it can happen that the upper level is more highly populated with electrons than the ground state. This process is called *population inversion*. The system is no longer in thermal equilibrium. A spontaneous emission process can occur, or an incident light wave of the correct wavelength can stimulate all the emission processes with the correct phases with a certain probability. As a result, the wave leaving the material will increase in intensity. It can be shown that at least three energy levels are required to produce a population inversion.

To transfer from an amplifier to an oscillator, one requires feedback. In the optical range,this is provided by optical cavities, whose dimensions are large by comparison with the wavelength. In the simplest case, the cavity consists of two planar mirrors, which are parallel to one another and separated by the cavity length. The light passes to and fro between the mirrors if it strikes them at right angles. A certain fraction of the radiation is lost at each reflection, because the reflectivity is always less than one. Further losses occur on account of diffraction, mirror imperfections, and so on. For an oscillator to result, the losses must be less than the gain. The cavity prolongs the dwell time of the photons in it and thus ensures complete inversion elimination.

Table 5. Laser Systems

Laser system	Active medium	Excitation	Typical length in cm	Output power in Watt		Wavelength in cm
				continuous	pulsed	
Gas	Liquid molecular gases metal vapors	Gas discharges chemical excitation	10 – 50	$10^{-3} - 10^{4}$	$10^{-3} - 10^{5}$	$10^{-4} - 10^{1}$
Liquid	Organic dyes in solvents	Flash tubes, laser light	5	10^{-1}	10^{4}	$4 \cdot 10^{-5} - 10^{5}$
Semiconductor	Semiconductor components doped with Zn or Se	Electric current	0,1	10^{-1}	10^{4}	$6 \cdot 10^{-4} - 10^{-3}$
Solid-state	Crystals and glasses doped with metal atoms or rare earths	Flash tubes, continuous gas-discharge lamps, tungsten strip lamps	2,5 – 20	$10^{-2} - 10^{-2}$	$10^{4} - 10^{-9}$	$5 \cdot 10^{-5} - 5 \cdot 10^{-2}$

A laser has three major components: an invertible material, an optical cavity, and an energy pump to produce the population inversion. A laser beam is characterized by the following features: high collimation (resonator property), monochromaticity (defined transition), spatial and time coherence (induced emission), high power density at the focus of a lens, and high frequency (optical transition).
There are many different laser systems. Typical data are provided in Table 6 ([31 a], supplemented):
The systems now known, which number more than 100, offer scope for producing about 1000 different wavelengths. See the literature for details.
In accordance with the mode of excitation, a distinction is drawn between *continuous-wave lasers* and *pulsed lasers*, and there is also a distinction between three-level and four-level systems in accordance with the number of energy levels involved in the process. Certain systems offer scope for shifting the wavelength over narrow bands, such as the dye laser.
Some typical laser applications are given below.

Table 6. Laser applications

Application	Property
Adjustment techniques	Spatial coherence
Separation/thickness measurement	Spatial coherence
Distance measurement	High intensity and spatial coherence
Interferometry	Time and spatial coherence
Holography	Time and spatial coherence
Material processing	High intensity, spatial coherence
Spectroscopy	Time coherence
Nonlinear effects	High intensity, coherence
Optical communications	High intensity, coherence

The laser effect was first demonstrated by the physicists Maiman and Townes in the USA and Basov and Prochorov in the USSR.

MAIMAN, Theodore Harold
American physicist, 11 July 1927 Los Angeles, industrial physicist at Hughes Research Laboratories in Miami

BASOV, Nikolai, Gennadevitch,
Russian physicist, 14 December 1922 Usman near Voronezh

PROCHOROV, Alexandr Michailowitsch,
Russian physicist, 11 Juliy 1916 Atherton (Australia), from 1958 Professor in Moscow

TOWNES, Charles Hard,
American physicist, 28 July 1915 Greenville (S.C.), Professor in New York, Cambridge (Mass.)

Grote, N., *Phys. Unserer Zeit* **8**, 103 (1977)
Induced Emission (1916):
Einstein, A., *Verh. Dtsch. Phys. Ges.* **18**, 318 (1916)
Einstein, A., *Phys. Z.* **18**, 121 (1917)
Lasers:
Ladenburg, R. and Kopfermann, H., *Z. Phys. Chem. Abt A* **139**, 375 (1928)
Purcell, E.M. and Pound, R.V., *Phys. Rev.* (2), **81**, 279 (1951)
Weber, J., *Trans. IRE PGED* **3**, 1 (1953)
Schalow, A.L. and Townes, C.H., *Phys. Rev.* **112**, 1940 (1958)
Basov, N.G. and Prochorov, A.M., *Zh. Eksp. Teor. Fiz.* **27**, 431 (1954)
Basov, N.G. and Prochorov, A.M. *Zh. Eksp. Teor. Fiz.* **28**, 249 (1955)
Maiman, T.H., *Nature* **187**, 493 (1960)
Lit.: [19, 33, 48, 56, 57, 67] (additional lit. see Table 31)

Lattice Peltier effect

→ thermoelectric effects

Laue effect (1912) SS

(Laue method)
X-ray spectroscopy began with the discovery of the Laue effect in 1912. Röntgen had recognized and described the important properties of his X-rays. However, a decision on the nature of X-rays was provided by Laue's experiment, in which the wave nature of the radiation was identified by using a crystal as a grating. The grating separation was a few times 10^{-8} cm, so the wavelength of the radiation should be of the same order. Together with Friedrich and Knipping, he demonstrated the interference of X-rays on zinc blende plates 0,5 cm thick. This led to the recognition of the wave nature of this radiation, in conjunction with the experiments of Barkla (1905) and others on the polarization of X radia-

tion and the experiments by Moseley (1913/1914) on the laws of discrete X radiation.
The occurrence of interference is called the Laue effect, and the detection method is known as the Laue method. It has remained one of the basic methods in the structure analysis of solids until the present day.
→ Bragg effect

LAUE, Max von,
German physicist, 9 October 1879 – 24 April 1960, Professor in Zürich, Frankfurt a.M., Berlin, and Göttingen

MOSELEY, Henry Gwyn Jeffrey,
British physicist, 23 November 1887 – 10 August 1915,

Laue, M. v., Knipping, P., and Friedrich, W., *Sitz.-Ber. der Kgl. Bayer. Akad. der Wissenschaften*, 303 (1912)
Röntgen, C.W., *Wiedem. Ann.* **64,** 1 (1898)
Barka, C.G., *Philos. Trans. London* **204,** 467 (1905)
Bragg, W.H. and Bragg, W.L., *Proc. R. Soc. London Ser. A* **88,** 428 (1913)
Lit.: [25, 36, 50 Vol. II, 2.2] (additional lit. see Table 51)

Laue method
→ Laue effect

Leidenfrost phenomenon (1756) Me
If a drop of water is placed on a very hot smooth surface, it does not evaporate at once because a poorly conducting layer of vapor is formed, which prevents contact with the surface and maintains the drop in the floating state. The effect is also observed with noble and heavy metals when these are brought into contact with water in the red-hot state, and the same occurs with dry ice.
→ Marangoni effect

LEIDENFROST, Johann Gottlieb,
German doctor, 27 November 1715 – 2 December 1794, from 1743 Professor in Duisburg
Lit.: [69] [76], *general textbooks*

Lenard effect (1890) Te, Ga, CC
(alternating current-direct current effect)
In a metal of noncubic structure, the DC resistance is greater than the AC value. This phenomenon occurs because of the production of a transverse temperature difference due to the transverse Peltier effect. A longitudinal thermal force is formed as a result of the transverse force. In the case of alternating current of sufficiently high frequency, the transverse temperature difference cannot occur, because a certain time is required for its formation. An analogous effect (the Lenard effect) occurs in all metals in a transverse magnetic field. Here, however, the DC resistance is smaller than the AC value.
→ thermoelectric effects

LENARD, Philipp Eduard Anton,
German physicist, 7 June 1862 – 20 May 1947, finally Professor in Heidelberg
Lenard, P.E.A., *Wiedem. Ann. Phys.* **39,** 28 (1890)
Lit.: [76] (additional lit. see Table 23, 30, 53)

Lenard effect (1892) EM
Water droplets are always charged due to the presence of molecular forces between water and the surrounding air. If the surface of a water droplet is removed, the result is a floating, negatively charged droplet, while the removed water is positively charged (*waterfall electricity* or *ballo electricity*, as called by Lenard, *spray effect*).
Lenard, P.E.A., *Wiedem. Ann.* **46,** 584 (1892)
Lit.: [76, 27a Vol. 4, part 1] (additional lit. see Table 26)

Length effect
→ piezoelectric effect
→ Zeeman effect

Length magnetostriction
→ magnetostriction

Lens effect L
The simultaneous irradiation of a crystal with an intense laser beam and an observation beam results in a lens effect of the observation beam produced by the pumping light.
→ nonlinear optical effects, → Kerr effect, → AC Kerr effect, → Raman effect
Hellwarth, R.W., *Progr. Quant. Electron* **5,** 1 (1977)
Lit.: [98, 100, 106]

Lichtenberg figures
→ Kirlian effect

Light deflection (1911) R

(gravitational effect, deflection effect)
Electromagnetic waves are deflected from a straight-line path in the gravitational field of a massive star. General relativity theory predicts a deflection of 1.75″ for light passing the margin of the Sun. The experimental values lie between 1.61″ and 2.73″. On account of the considerable experimental difficulties, these values provide a confirmation of the theory. The first such calculation was made by Soldner in 1801 on the basis of Newton's theory of gravitation, which gave half the above value (this paper is reprinted in *Ann. Phys.* **65**, 593 (1921)). In 1911, Einstein arrived at Soldner's value.
→ Shapiro experiments (1967)
Einstein, A., *Ann. Phys.* (4) **35**, 898 (1911)
Lit.: [6c, 59] (additional lit. see Table 43)

Light-electric effect

→ photo effect, → Hallwachs effect

Light-hydraulic effect

When an intense pulsed laser beam passes through a liquid, shock waves can occur, whose pressures can be employed in processing and hardening materials.
→ nonlinear optical effects
Lit.: [75]

Limb effect

→ gravitational frequency shift

Linear Doppler effect

→ optical Doppler effect

Linear electrooptic effect

→ Pockels effect

Liquid crystal effects LC

Liquid crystals have been known for almost a hundred years. For a long time, there was very little interest in substances that were neither liquid nor crystal nor solid. A liquid crystal behaves like a liquid in assuming the shape of the container, whereas on the other hand it also exhibits typically crystalline behavior. Liquid-crystal compounds are always organic. They have extended planar molecular structures, which are rigid to a certain extent. The long, rod-shaped molecules either have large dipole moments or are readily polarized. A distinction is drawn between cholesterolic, smectic, and nematic liquid crystals. Thermooptic and electrooptic effects can be observed in liquid crystals. The electrooptic effects of nematic liquid crystals are of particular interest. Wide use is made of liquid crystals in displays for watches and desk calculators.
Sobel, A. *Sci. Am.* **228**, 65 (1973)
Lit.: [100, 114]

Bistability effect

(hysteresis effect)
In the bistability effect, which occurs in special cholesterolic liquid crystals, there is switching between the cholesterolic and nematic homotropic textures, as in texture switching. Hysteresis occurs, accompanied by bistable behavior.

Electrohydrodynamic effects

Electrodynamic effects in nematic liquids are linked to current flow and mechanical liquid flow. In dynamic light scattering, the oriented nematic liquid crystal layer becomes light scattering in an electric field on account of hydrodynamic turbulence.

Field effects

The dielectric constant of a liquid crystal shows marked anisotropy: the dipole moment lies either in the direction of the molecular axis or at a certain angle to it. An external electric field then exerts a torque on the molecule, and, if there is positive anisotropy in the dielectric constant, the molecules are aligned in the field direction. The effect is visible also in thin layers between polarizers because of the high birefringence or the rotation of the direction of polarization through 90°.

Guest-host effect

If dye molecules that have the property of absorbing the light in one polarization direction are dissolved in a liquid crystal, the dye molecules align themselves with the liquid-crystal ones. As the absorption spectrum of these molecules is a function of orientation and of polarization direction in the incident light, their orientation determines whether the characteristic color of the dye appears or white light.

Memory effect
In the *Heilmeier memory effect*, a cholesterolic liquid-crystal layer becomes light scattering on account of electrohydrodynamic turbulence. This state is attained above a certain critical field strength. The cholesterolic texture is transparent, with the axis of the crystal perpendicular to the layer plane. The electric field rotates this axis. The resulting texture is called *focal-conical*. It is strongly birefringent and inhomogeneous. This leads to light scattering and depolarization. The focal-conical state is metastable and can persist for long periods in the field-free state. It can be erased by means of a high-frequency field.

OPD effect
In oriented phase deformation (OPD) the long axes in a liquid crystal molecule with negative anisotropy in the dielectric constant are originally aligned perpendicular to the electrode surfaces. The liquid crystal orients spontaneously due to wall forces and intermolecular elastic forces. An electric field exerts an additional electric torque on the molecules. When a certain field strength is exceeded, the initial orientation is deformed. The transmission of the layer between crossed polarizers is a function of the field strength, which controls the deformation angle (variable birefringence).
OPD cells allow voltage-dependent colors to be produced.

Schadt-Helfrich effect (1971)
In the Schadt-Helfrich effect, a nematic liquid crystal layer exhibiting positive anisotropy in the dielectric constant is oriented parallel to the boundary surfaces. The arrangement is chosen so that the molecules in the liquid crystal layer are twisted through 90°. Linearly polarized light passing through the layer also shows rotation of the plane of polarization through 90°. In the absence of an external field, such a layer placed between crossed (parallel) polarizers is transparent (opaque).
In a field, the layer between crossed (parallel) polarizers becomes opaque (transparent).
Schadt, M. and Helfrich, W., *Appl. Phys. Lett.* **18,** 127 (1971)

Texture conversion effect
This is a special case of the → memory effect in a cholesterolic liquid crystal with positive anisotropy in the dielectric constant. In this effect, there is switching back and forth between cholesterolic (focal-conical) and nematic textures. The fact is utilized that a suitable bias near the threshold voltage reduces the switch-on time yet increases the switch-off time.

Thermooptic effect
A change in the optical properties of a material occurs due to heat radiation. Effects of this type are particularly pronounced in liquid crystals and can be used as a heat detector.
Lit.: [29d]

Longitudinal barrier-layer photoeffect
→ barrier layer photoeffect

Longitudinal Doppler effect
→ optical Doppler effect

Longitudinal effect
→ braking effect

Lossew effect (1923) Lu
Lossew observed luminescence from an SiC rectifier in the neighborhood of the contacts, with differences in emission at the anode and cathode. This arises from injection luminescence in the conducting direction and collisional ionization in the nonconducting direction.
Lit.: [6d], dictionaries

Ludwig-Soret effect
→ diffusion thermoeffects

Luminescence effects Lu
Luminescence is the emission of light from a material previously stimulated by energy absorption. Wiedemann introduced the term in 1889. It covers the phenomena of *fluorescence* and *phosphorescence*, which differ physically only in their decay times or sometimes in their emission mechanisms. The expressions are now becoming more and more dispensable. In the following table luminescence effects are distinguished in accordance with the mode of excitation.

Table 7. Luminescence effects

Name	Excited by
Photoluminescence	optical radiation
Radioluminescence	ionizing radiation
Cathodoluminescence	electrons
Chemiluminescence	elementary processes of a chemical reaction
Bioluminescence	reactions occuring within organisms
Electroluminescence	electric fields
Triboluminescence	mechanical processes
Thermoluminescence	based on previous excitation at low temperatures, followed by luminescence stimulation by raising the temperature

Electroluminescence has considerable technical significance (previously cathodoluminescence was sometimes also included under this term. It provides the basis for flat picture production (Destriau effect), light intensification (image intensifiers), and particularly in the excitation of light-emitting diodes (LED) and laser diodes (optoelectronics).
Electroluminescence is ascribed to the following mechanisms: 1. Charge carrier injection; 2. Collisional ionization in crystals; 3. Internal field emission.
Lit.: [6d, 8, 17, 22d, 27, 30, 49, 72 – 78, 90]

Macaluso-Corbino effect (1898) Mo

This is the anomalously large magnetic rotation of the plane of polarization in the neighborhood of absorption lines, as observed in low-density gases, for which the normal rotation is immeasurably small.
magnetooptic effects

MACALUSO, Danviano.
Italian physicist, 1845 – ?

CORBINO: → Corbino effect

Macaluso, D. and Corbino, O.M., *C.R. Acord. Sci. Paris* **127**, 548 (1898)
Corbino, O.M., *Nuovo Cimento* **8**, 257 (1898)
Corbino, O.M., *Nuovo Cimento* **8**, 381 (1899)
Lit.: [50 Vol. II, 2.2, 75, 77] (additional lit. see Table 37)

Macroscopic quantum effects

→ superconduction, → Onnes effects

Maggi effect

(external photoelectric effect)
→ photoeffects

Maggi-Righi effect

→ thermomagnetic effects

Magnetic aftereffects G, EM

Magnetic aftereffects include the *thermal aftereffect* (also called the *Jordan aftereffect*) and *diffusion aftereffects*, including the *Richter aftereffect*.
In the Jordan aftereffect, there are thermodynamic fluctuations in the spin system and hence in the magnetic energy of a ferromagnetic material. The effect resembles Brownian motion in producing statistically varying local field strengths within the magnetic domains and therefore thermal Barkhausen steps. The Jordan aftereffect is controlled by the external magnetic field and facilitates its action: the magnetization attains higher levels than without the effect. The Jordan aftereffect is related to hysteresis phenomena and occurs in all magnetic materials.
Diffusion aftereffects are related to site-exchange processes in a material: the magnetization structures interact with the lattice components and lattice defects, which can diffuse through the crystal.
Every magnetization structure favors a certain arrangement, e.g., of the interstitial atoms, which can be altered more readily if the interstitial atoms have time to follow the changes. The permeability thus has a smaller value at high frequencies than it does at lower frequencies, where the aftereffect makes itself felt. A classic example of a diffusion aftereffect is that of carbon atoms in iron (→ Snoek effect), which is called the Richter aftereffect. The effect was examined in detail by Ritter and was classified by Snoek.
In ferrites, electron diffusion, called the Wijn aftereffect, as well as cation diffusion are found. Other magnetic aftereffects are spin relaxation and eddy-current aftereffects.
Lit.: [18, 31, 37] (additional lit. see Table 19 and 26)

Magnetic birefringence

→ magnetooptic effects, → Cotton-Mouton effect

Magnetic bubbles

→ Bobeck effect

Magnetic contraction effect Mm

(inverse magnetostriction effect)
If a ferromagnetic material is subjected to tension, its remanence alters. The tension affects the spontaneous distribution of the Weiss domains. If the magnetostriction is positive, the remanence increases, while if it is negative, it decreases.
→ magnetostriction
Lit.: [4, 34, 50 Vol. 4, part 4] [76] (additional lit. see Table 36)

Magnetic memory effect

→ Bobeck effect

Magnetic resistance effect

→ galvanomagnetic effects

Magnetic rotation of the plane of polarization

→ Faraday effect

Magnetoacoustic effect SS

In the absorption of ultrasound in single metallic crystals, oscillations in absorption, which are dependent on the field strength, occur at low temperatures in a homogeneous magnetic field. The effect is largest when the magnetic field is perpendicular to the direction of ultrasound propagation. The electrons move over the Fermi surface (→ de Haas-van Alphen effect) with the cyclotron frequency, along circular paths controlled by the magnetic field. The ultrasonic wave perturbs the lattice and produces periodic electric fields. If the periodic electric field is in resonance with the cyclotron frequency, maximal energy transfer takes place. The absorption is then strongest if the electrons move parallel to the phase surfaces of the ultrasonic wave. Perpendicular to this there is minimal absorption.
The Fermi surface in the material can be evaluated by means of the magnetoacoustic effect.
Lit.: [1, 71, 75, 76] (additional lit. see Table 51)

Magnetocaloric effect (1918) LT

When thermally insulated certain paramagnetic substances show sudden changes in magnetization in response to temperature change – an effect which is used in producing temperatures slightly above absolute zero in low-temperature physics. The magnetic field is suddenly switched off, which produces a temperature reduction in the sample.
→ electrocaloric effect
Weiss, P. and Picard, A., *C.R.* **116,** 352 (1918)
Lit.: [3, 26] (additional lit. see Table 33)

Magnetoelasticity effect

→ ΔE effect, → magnetomechanical effects

Magnetoelectric effect EM, Ga

In insulating materials with certain complicated structures related to magnetic ordering, an electric field produces proportional magnetization, while a magnetic field produces proportional elecrical polarization. The explanation is based on exchange effects and spin-path interaction in the material. The effects in chromium (III) oxide have been examined.
Lit.: [76] (additional lit. see Table 26 and 30)

Magnetohydrodynamic effect Pl

A homogeneous magnetic field applied to a compressible conducting liquid produces a voltage when the liquid moves. Further, this produces electric currents, which result in a change in the magnetic field. The interaction between the currents and the magnetic field produces forces which influence the state of motion. In response to a wave perturbation, the electric field, the magnetic field, and the current, as well as the hydrodynamic pressure, propagate as waves in the direction of the magnetic field. These waves are called magnetohydrodynamic or *Alfvén waves*. The Alfvén waves play an important role in explaining certain phenomena in astrophysics. In the *magnetohydrodynamic generator* use is made of charge transport in a conducting fluid in the presence of a magnetic field to produce current. In principle, this is a demonstration of the → Hall effect in a rapidly flowing ionized gas in a transverse magnetic field. This separates the electrons and ions. The voltage is proportional to the magnetic field, the flow speed, and the electrode separation, while the current density is proportional to the magnetic field and the flow speed.
All magnetohydrodynamic effects are due to the Lorentz force.
ALFVÉN, Hannes,
Swedish physicist, 1909
Lit.: [6e, 43, 76] (additional lit. see Table 41)

Magnetomechanical effects Mm

(magnetostriction effects, → magnetoelastic effects)

In this group of effects, there is coupling between the mechanical and magnetic properties of ferrimagnetic or ferromagnetic materials. Either the mechanical or the magnetic properties can alter.

Two large groups are distinguished:

1. A magnetic field produces mechanical changes in the sample, for example, magnetostriction or the Wiedemann effect.
2. Mechanical forces produce magnetic effects, for example, the Villari effect and the Wertheim effect.

The following relationship exists between magnetic-elastic and elastic-magnetic effects: when *tension* facilitates (hinders) magnetization, magnetization should produce an extension (contraction of the sample). When *pressure* facilitates (hinders) magnetization, magnetization should produce a contraction (extension) of the sample. If the volume is increased (decreased) by magnetization, a hydrostatic tension (pressure) increases the magnetization, and vice versa.

→ magnetostriction, → Δ *E* effect, → Villari effect, → Matteucci effect, → magnetic tension effect, → Wiedemann effect

Lit.: [18, 50 Vol. 4, part 4] (additional lit. see Table 36)

Magnetooptic effects Mo

As in electrooptic birefringence, there are two experimental arrangements: the magnetic field is either transverse or longitudinal to the light propagation direction. When light passes through an optically transparent medium, there may be birefringence or rotation of the plane of polarization, or line splitting if the radiation source is in the magnetic field. The classical explanation for the effects is provided either by the anisotropy produced by the magnetic field or by the Lorentz force. The effects can be observed with light and microwaves (Table 8).

The polarization change on reflection at a magnetic mirror is known as the → magnetooptic Kerr effect.

The magnetooptic effects are of importance in the examination of solids, particularly semiconductors. In conjunction with current-conduction effects, they give evidence of the internal structure, such as band separations, etc.

Lit.: [6c, 9, 18, 22c, 27, 50 Vol. II, 2.2] (additional lit. see Table 37)

Table 8. Magnetooptic Effects

Action	Change in polarization direction or refractive index	Line splitting
Magnetic field transverse to propagation direction	Cotton-Mouton effect Voigt effect Majorana effect Hanle effect (decrease in polarization)	Zeeman effect, three components (transverse effect) Paschen-Back effect (transition or transformation effect)
Magnetic field longitudinal to propagation direction	Faraday effect Becquerel effect Kundt effect Maculoso-Corbino effect	Zeeman effect two components (longitudinal effect)

Magnetooptic Kerr effect (1876) Mo

In general, elliptically polarized light is produced when polarized light is reflected at a metal mirror. The changes in phase and amplitude in the reflected light occurring when a ferromagnetic mirror is highly magnetized are called the magnetooptic Kerr effect; if linearly polarized light is reflected at the polished surface of a magnetized material, the plane of polarization is rotated (the smooth surface of a thin film may also be used,

and the magnetization may lie in the surface plane).
This effect can be used, for instance, to examine the states of magnetization in metal surfaces and the changes occurring in the magnetic domains, or Weiss areas. The effect is closely related to the → Kundt effect. A distinction is made between *longitudinal, transverse*, and *polar magnetooptic Kerr effects*.
→ magnetooptic effects

KERR, J.,
17 April 1824 – 1907
Kerr, J., *Rep. Brit. Ass.* 85 (1876)
Kerr, J., *Philos Mag.* (5), **3**, 321 (1877); **5**, 161 (1878)
Lit.: [26, 34, 50 Vol. II, 2.2] (additional lit. see Table 37)

Magnetooptic transformation
→ Paschen-Back effect

Magnetoplasma effects SC
This is the collective term for effects produced by a magnetic field in a semiconductor plasma. For example, changes take place in the excitation of electromagnetic waves in a semiconductor plasma when the magnetic field is in the wave propagation direction. An explanation for the effects is provided by the Lorentz force and cyclotron resonance of the electrons in the semiconductor plasma.
Lit.: [36]

Magnetoresistance
→ galvanomagnetic effects

Magnetorotation
→ Faraday effect

Magnetostriction (1842) Mm
(Joule effect)
Magnetostriction is the name given to any change in the geometrical dimensions of a body produced by a change in magnetization. A distinction is made between shape magnetostriction and volume magnetostriction.
The magnetically dependent volume-invariant form change is often called the Joule effect, Joule magnetostriction, or magnetostriction in the narrow sense. The relative change in length between the demagnetized state and the magnetically saturated state (saturation magnetostriction) lies between 0 and 10^{-4}. One speaks of positive (or negative) magnetostriction when a magnetic field produces an increase (or decrease) in length in the direction of the magnetic field. The converse (inverse magnetostriction) is the magnetic tension effect.
In volume magnetostriction, the changes in length are one to two orders of magnitude smaller than those in form magnetostriction. One cause of volume magnetostriction is the shape effect, in which the magnetostatic energy of a body is dependent on its volume in its own demagnetizing field. Magnetostriction is used to produce ultrasound and in fine adjustment, e.g. of laser mirrors.
→ magnetomechanical effects
Joule, J.P., *Ann. Electr. Magn. Chem.* **8**, 219 (1842)
Lit.: [3, 34, 50 Vol. 4, part 4] [92] (additional lit. see Table 36)

Magnus effect (1852) F
A body rotating in a flowing gas or liquid with its axis of rotation perpendicular to the flow experiences a force perpendicular to both directions. The effect plays an important part in ballistics (projectile spin, sliced tennis ball).
Attempts have been made to use the effect for propelling ships and aircraft, by replacing the sail by a rotating tower (Flettner rotor).

MAGNUS, Heinrich Gustav,
German physicist and chemist, 2 May 1802 – 4 April 1870, from 1834 Professor in Berlin
Ackeret, J.: *The Rotor Ship and its Physical Basis* [in German]. Vandenhoeck & Ruprecht, Göttingen 1925
Swanson, W.M., *ASME, J. Basic Eng.* **83**, 461 (1961)
Lit.: [6a, 69, 2nd Ed.] (additional lit. see Table 29)

Majorana effect (1902) Mo
A transverse magnetic field produces optical anisotropy in a colloidal solution such as a sol of iron oxide, which leads to magnetic birefringence.
→ magnetooptic effects

MAJORANA, Quirino,
Italian physicist, 28 October 1871 – 31 July 1957
Majorana, Qu., *Rend. Lincei* (5) **11**, 1, 374
Majorana, Qu. *Phys. Z.* **4**, 145 (1902)
Majorana, Qu. *Rend. Lincei* (5) **2**, 90, 139 (1902)
Lit.: [22c, 50 Vol. II, 2.2, 89] (additional lit. see Table 37)

Malus's experiment (1808 – 1811) O

If light is reflected from a glass plate and the reflected beam falls on a second glass plate, the light intensity can be varied by rotating the second glass plate about the axis of the first reflected beam, and under certain conditions the intensity can become zero (Malus 1808). The two glass plates act as polarizer and analyzer, and partially polarized light is produced by reflection, the component oscillating perpendicular to the plane of incidence being reflected preferentially. This applies also for the second glass plate. If the angle of incidence is equal to the Brewster angle, linearly polarized light results. The *Brewster angle* is defined as the angle at which the tangent of the angle of incidence is equal to the refractive index. For glass ($n \approx 1.5$), the polarization angle is $\alpha \approx 57°$.

MALUS, Etienne Louis,
French engineer and physicist, 23 June 1775 – 23 February 1812, from 1811 teacher at the Ecole Polytechnique in Paris
Laplace, P.S., *Ann. Phys.* **31**, 274 (1809)
Lit.: [6c, 9, 22c, 50 Vol. II, 1] (additional lit. see Table 39)

Marangoni effect Me

The fact that under certain conditions water droplets can dance on a water surface was already examined by J. Tyndall in 1885.

The lifetime of the drops on the water surface is dependent on many parameters: surface tension, purity, and state of motion of the surface. A relationship to the → Leidenfrost phenomenon is likely, assuming that small air cushions exist under the drops. The drops sink when the cushions disappear. If the surface tension under a droplet is reduced, the layers near the surface displace the air cushion and the drops move in a certain direction until they sink. This is called the Marangoni effect. An understanding of this phenomenon is important in elucidating fog formation and dissipation.

Marangoni, C., *Nuovo Cimento* **2**, 5, 239 (1871); **3**, 50, 97 (1878)
Shaplanol, D., and Rycroft, M.: *Spacelab.* Cambridge University Press, Cambridge – London – New York 1984

Masers

→ lasers

Masking effect Me

The ear is less sensitive to high frequencies if it is also supplied with a tone or narrow-band noise at a low frequency. The effect becomes more pronounced the stronger the masking tone and the closer the masked frequencies lie to this tone.
Lit.: [76, 85], *dictionaries, handbooks*

Mass effect

→ analogue states

Matteucci effect (1847) Mm

If an iron rod is twisted in a longitudinal magnetic field, its magnetization alters. Twisting and untwisting of the magnetic rod leads to induction currents. Thomson showed that slight hysteresis is associated with the effect.

In addition to this *positive* effect, Schmoller found a *negative Matteucci effect*, in which increasing magnetization in the twisted rod occurs together with oppositely directed electromotive forces.

MATTEUCCI, Carlo,
Italian physicist, 20 June 1811 – 25 June 1868, Professor in Pisa
Matteucci, C., *Ann. Chim. Phys.* **5**, 416 (1858)
Schmoller, F.Z., *Phys.* **93**, 35 (1935)
Lit.: [26, 50 Vol. 4, part 4] (additional lit. see Table 36)

Maxwell effect O

In a flowing liquid optical anisotropy occurs in the presence of velocity gradients. If the fluid rotates, the Coriolis force can produce effects which statistically will cancel each other. For this reason, the only effects that persist are those associated with linear motion of the fluid. Suspensions and colloidal solutions of anisotropic particles acquire orientation because of velocity gradients in the flow. This is called the Maxwell effect.

MAXWELL, James Clark,
Scottish physicist, 13 November 1831 – 5 November 1879
Lit.: [28 Vol. VIII] (additional lit. see Table 39)

Maxwell-Lippmann effect

→ shielding effect

Meissner effect S

(Meissner-Ochsenfeld effect, displacement effect)

Lippmann's phenomenological theory of superconductors (1919) treats these as conductors with an infinitely great electrical conductivity. Maxwell's electrodynamics then implies an important consequence, namely, that the inner part of a superconductor is electromagnetically completely screened from the outer space. If a superconductor is introduced into a static magnetic field, the inner part of the conductor remains field free. This has been demonstrated by direct field measurements. However, the result of this experiment should differ if a body is introduced into the magnetic field above the transition temperature and is then made superconducting by cooling. In that case, the magnetic field should pass unperturbed through a superconductor. Experiment (Meissner 1933) decisively contradicted this prediction, in that it showed that the interior of a suitably thick superconductor always remains magnetic-field free in an external magnetic field; apart from a thin marginal layer, the magnetic field is ejected from the superconductor. The screening of the magnetic field is provided by a persistent current flowing in the boundary layer. This is proportional to the flux change occurring in the conductor.

MEISSNER, Fritz Walther,
16 December 1882 – 16 November 1974, from 1934 Professor in Munich

Meissner, R. W. and Ochsenfeld, R., *Naturwissenschaften,* **21,** 787 (1933)

Lit.: [3, 6b, 6d, 11, 15, 21, 22b, 22d, 27, 30, 38 Vol. V, VI, VIII, 41, 70 Vol. II, 74, 75, 90] (additional lit. see Table 44)

Meissner-Ochsenfeld effect

→ Meissner effect

Memory effect (1972) SS

(shape memory effect SME)

Certain alloy systems have the property of remembering their mechanical states at a certain temperature. Bending at low temperatures leads, for example, to straightening at high temperatures, with return to the bent form on cooling. The process is reversible and can be used as a switch, motor, or drive.

The effect is explained in terms of martensite and similar transformations. Minor movements of the atomic domains are thought to be transmitted to the complete set of atoms, which affects the mechanical properties.

A distinction is drawn between the elastic form memory effect (*superelasticity* or *ferroelasticity*), the *plastic SME* (in the martensite phase), and the *two-way SME* (in the austenite phase), which is incomplete and may be eliminated. There are various other memory effects, such as the → relaxation effects, → Ovshinsky effects, etc.

Table 9. Memory alloys

Copper-tin
Copper-zinc
Copper-aluminum
Copper-aluminum-nickel
Manganese-copper (5 – 50 %)
Gold-cadmium
Cadmium-silver-gold
Nickel-titanium (50 – 60 %)
Iron-platinum
Indium-thallium
Uranium-molybdenum (2 – 7 %)
Uranium-niobium (3 – 11 %)
Uranium-rhenium (2 – 7 %)

Tas et al., *J. Less-Common Met.,* **28,** 141 (1972)
Redakt.-Ber., Phys. Unserer Zeit, **8,** 33 (1977)
Robinson, A. L., *Science,* **191,** 934 (1976)
Lit.: [28a] (additional lit. see Table 51)

Memory effect

→ liquid-crystal effects

Merrington effect F

Any fluid exhibiting abnormal flow behavior (*non-Newtonian flow, Reiner-Rivlin fluids*) shows pronounced expansion on account of the positive normal stresses on flowing out of a nozzle, where the diameter of the fluid flowing out can be several times the nozzle diameter. The expansion increases with the flow rate and as the nozzle diameter becomes smaller.

Lit.: [75] (additional lit. see Table 29)

Mesomeric effect

→substituent effects

Michelson's experiment (1881) R

The aim of this experiment was to measure the speed of the Earth relative to the ether at rest. Michelson's experiment conflicted with the hypothesis that the ether exists. The negative result of the experiment led to the conclusion that the velocity of light in a vacuum is constant in all directions.

In a *Michelson interferometer,* a parallel beam of light is split into two partial beams perpendicular to one another, which are reflected at two mirrors after travelling equal distances. The partial beams are then recombined, and one observes the interference.

In accordance with the ether theory, light moving in the direction of motion of the Earth should differ in speed from that moving perpendicular to it, so that when the apparatus is turned through 90° a shift should be observed in the interference fringes. Since the resolution of the interferometer is 10^{-15}, and the calculation for the Earth in motion (rotation about the Sun) gives 10^{-8}, a distinct effect should be observed.

In fact, however, no shift is observable in the interference fringes. Other experiments rule out the ether being transported with the Earth, so the ether hypothesis must be abandoned.

Michelson's experiment was performed with the American chemist Eduard Williams Morley (1838 – 1923) in 1887, and was subsequently repeated with increasing precision. The use of the → Mössbauer effect by Champeney (1963) led to the same result.

MICHELSON, Albert Abraham,
American physicist of Polish extraction, 19 December 1852 – 9 May 1931, finally Professor in Chicago

Michelson, A. A., *Astrophys. J.* **61,** 9 (1925)
Michelson, A. A., *Am J. Sci.* **22,** 6 (1881)
Michelson, A. A. *Am. J. Sci.* **31,** Vol. 77 (1886)
Michelson, A. A. and Morley, E. W., *Am J. Sci* (3) 333 (1887)
Michelson, A. A., *Phil. Mag.* (5) **24,** 463 (1887)
Lit.: [6c, 10, 16, 22c, 40 Vol. I, 50 Vol. 5, 2, 61 Vol. 4], *dictionaries* (additional lit. see Table 43)

Mie effect (1908) Sc

(Mie scattering)

The Mie effect occurs when light is scattered at particles whose dimensions are comparable with or greater than the wavelength. There is an increase in the back-scattering as the radius of the scattering particles increases.

The scattered waves from the different areas in a scattering particle undergo phase shifts, and will thus cancel out due to interference in certain directions, whereas in others they will be intensified. The effect plays a part in light scattering at dust particles in the atmosphere. It is found that the scattering of large particles in the atmosphere is described better by Mie scattering than by → Rayleigh scattering.

→ scattering effects

MIE, Gustav Adolf Ludwig,
German physicist, 29 September 1868 – 13 February 1957, finally Professor in Freiburg im Breisgau

Lit.: [6c, 9, 74, 75, 93] (additional lit. see Table 45)

Modal noise effect (1978)

(modal distortion, modal noise)

An interfering effect occurs in multimode fibers if these are driven by a light source with a large coherent length. The light intensity in the fiber is spatially overlapped by a granulation pattern (speckle pattern), which alters if:

1. The fibre is in motion,
2. The laser spectrum alters on account of temperature fluctuation, working-point shift, or as a result of modulation.

The mode-dependent attenuation in connectors, splices, or in the fibre itself causes the speckle pattern to be converted to time-dependent fluctuations in the light power received by the photocell. This results in noise or signal interference.

Epworth, R. E., in: *Fourth European Conference on Optical Communication.* Genoa, September 1978
Standard Telecommunication Laboratories Limited, Harlow, Essex

Møller scattering

→ particle scattering, → Mott scattering

Mössbauer effect (1958) N

(resonant γ-ray absorption)

The Mössbauer effect leads to an understanding of the absorption and emission of gamma rays by atomic nuclei. With gaseous gamma sources, the shape of the emitted spectrum is affected by the

thermal motion of the molecules, the recoil occurring on emission, and the Doppler effect. If the atoms are incorporated into suitable crystals, the nuclei can be made to emit virtually recoil-free gamma rays at low temperatures (the Mössbauer effect). The natural line width in emission can be observed and no Doppler broadening occurs.
The Mössbauer effect allows one to make energy and frequency measurements with an uncertainty of 10^{-15}. The Mössbauer effect confirms the constancy of the velocity of light and the relativistic formulas for the dependence of time and mass on velocity. It can also be used to test Einstein's general theory of relativity (red shift of the light frequency in a gravitational field).
Mössbauer observed the effect by experiment in 1958 and gave a theoretical interpretation.

MÖSSBAUER, Rudolf Ludwig,
German physicist, 31 January 1929 in Munich, 1961 – 1964 Professor at the California Institute of Technology in Pasadena, then in Munich, and Director of the Laue-Langevin Institute in Grenoble 1973 – 1978

Mössbauer, R. L., *Z. Phys.* **151**, 124 (1958)
Mössbauer, R. L., *Naturwissenschaften* **45**, 538 (1958)
Mössbauer, R. L., *Phys. Bl.* **18**, 97 (1962)
Lit.: [6c, 6d, 74, 75, 817] (additional lit. see Table 38)

Mott scattering
→ particle scattering

Nernst effect
→ galvanomagnetic effects

Neugebauer effect Eo
Neugebauer suggested a contribution to the change in polarization related to the electrooptic birefringence in an external electric field, but it is extremely small.
→ electrooptic effects

NEUGEBAUER, Theobald Ludwig,
Hungarian physicist, 30 May 1904

Lit.: [76] (additional lit. see Table 27)

Nikischov effect AQ II
The production of electron-positron pairs by the collision of low-energy photons with high-energy photons (γ-rays) is known as the Nikischov effect. The threshold for this process is very high, above 7.1 GeV.

Lit.: [14] (additional lit. see Table 21)

Noise effect
→ shot effect

Nonlinear-optical effects L
Under normal conditions, electrical polarization is linearly dependent on the electric field or the dielectric displacement. Virtually all effects in classical optics and crystal optics can be explained within the framework of this relationship.
However, in general higher terms in the series expansion affect polarization as a function of field strength; in addition to the linear term, effects from the quadratic and cubic terms are observed at high field strengths. These comprise the field of nonlinear optics. There are various effects, which in part are known also in acoustics and high-frequency engineering:

Table 10. Nonlinear optical effects

Name	Significance
Frequency doubling	Numerous crystals double the frequency of light passing through them
Optical mixing	If two light beams differing in frequency pass through a crystal together, one gets not merely the doubled frequencies but also the sum and difference frequencies
Optical rectification	Certain crystals act as rectifiers, i.e., they convert part of the AC field of the light into a DC field
Optical Kerr effect	A strong, linearly polarized field of light produces anisotropy in an isotropic medium, i.e., birefringence
Intensity-dependent transmission	The transmission of an absorbing medium can alter in response to intense light (passive optical switches)
Self-focussing	At high light intensities, the refractive index of a medium can alter in such a way that a light beam focusses itself

Name	Significance
Induced scattering	Normal light scattering occurs more or less equally in all spatial directions. Induced scattering occurs in a preferred direction, and the scattered light can be strengthened by induced effects
Ionization in gases	At extremely high light intensities, atoms can be ionized in the field of the beam

Lit.: [19, 22c, 32, 48, 56, 57, 67] (additional lit. see Table 31)

Normal photoelectric effect
→ photoelectric effects

Nuclear Hanle effect N
This is the dependence of the linear polarization of γ rays on the hyperfine interaction in an external magnetic field. The effect is equivalent to the → Hanle effect. Two cases may be distinguished:

1. The hyperfine splitting is substantially larger than the natural linewidth of the excited nuclear level. One finds right-handed and left-handed circularly polarized light, and no linear polarization.
2. The splitting is substantially less than the natural linewidth. Then the right-handed and left-handed circularly polarized components cannot be distinguished, and they interfere, so linearly polarized radiation results. The Zeeman splitting increases with the external magnetic field, and the degree of linear polarization decreases.

Heusinger, R. et. al., *Phys. Lett.* **49B**, 269 (1974)
Kreische, W. et al., *Phys. Rev.* **C 17**, 2006 (1978)
Lit.: [100, 114]

Nuclear magnetic (electrodynamic) resonance effects N
The interaction between the nuclear magnetic moment and electromagnetic waves leads to resonance effects, and the paramagnetic susceptibility related to the magnetic moment can be measured.

Nuclear magnetic absorption and dispersion are further examples of this type of effect.
Lit.: [76] (additional lit. see Table 38)

Nuclear photoeffect (1934) N
(Chadwick and Goldhaber 1934)
In the nuclear photoeffect, one or more neutrons may be emitted due to the absorption of gamma quanta. This usually leaves a radioactive nuclide. The nuclear photoeffect sets in above 5 MeV and has a cross section which is by one or two orders of magnitude smaller than those for other particle reactions. The effect was first observed with neutrons. The nuclear photoeffect can be used to produce almost monochromatic neutrons of medium energy.

GOLDHABER, Maurice,
18 April 1911 Lemberg

CHADWICK, James,
20 October 1891 – 23 July 1974
Chadwick, J. and Goldhaber, M., *Nature* **134**, 237 (1934)
Lit.: [96, 100, 114]

Nuclear resonance scattering
→ scattering effects

Nuclear spin effects N
Atomic nuclei possess an inherent angular momentum or spin. This can be influenced by electromagnetic fields, as can the spin of the shell electrons. In a magnetic field the spectrum shows a hyperfine structure produced by the nuclear spin. The energy terms can be calculated in the same way as for optical spectra (→ Zeeman effect etc.).
Lit.: [90] (additional lit. see Table 38)

Nuclear Thomson scattering
→ scattering effects

Nucleation effect (1943) S
In magnetically induced transitions from the superconducting state to the normal state, superconducting areas persist, even above the critical magnetic field strength. These areas act as nuclei, so it is possible for the transition to the superconducting state to occur without cooling below the transition temperature.
Lit.: [30] (additional lit. see Table 44)

Number effect
→ red shift

OEN effect

→ thermomagnetic effects

Onnes effects (1913) LT

(Kammerlingh Onnes, superfluidity)
The Onnes effect is the property of liquid helium II (isotope ^{4}He I) of passing over a barrier, e.g., over the edge of a beaker, as a thin film. The effect occurs if the temperature at the entire edge is less than 2.17 K. Below this temperature, all the properties of liquid helium change suddenly: the thermal conductivity becomes infinitely great, and the viscosity infinitely small. In addition, there are changes in density, specific heat, dielectric constant, speed of sound, and behavior under rotation.
The helium nucleus has spin 0, which leads to degeneration in the Bose-Einstein statistics at low temperatures, namely the *Bose-Einstein condensation* in momentum space (Einstein 1924). In this case, there are many atoms that only possess the zero-point energy, and therefore cannot transfer momentum. In simple terms, helium II can be thought of as consisting of two liquid components: first, that composed of particles having finite momentum, and second, that composed of particles lacking momentum. The superfluid component then has zero entropy, which corresponds to an absolute tempeature of 0 K. The superfluid particles free from momentum can move along the walls and through the other fluid particles without friction.
In flow through a capillary, there is transport of cold because of the lack of thermal energy. The *mechanocaloric effect* is that the container thereby becomes warmed, as the entropy in the container must increase. The law of conservation of energy and the second law of thermodynamics then imply that the pressure in the warm vessel must rise (the *thermodynamic pressure effect* or *the fountain effect*).
Lit.: [6] (additional lit. see Table 33)

Optical Doppler effect O

The phenomenon is similar to that of the acoustic Doppler effect. A frequency or wavelength shift is observed in light when the source is in motion: there is a blue shift if the source is moving towards the observer; and a → red shift if it is moving away from the observer. The optical Doppler effect is relativistic. In 1905, Stark discovered the optical Doppler effect in light from canal rays, where it is readily observed because of the high particle speed (*Stark-Doppler effect,* 1906).
The special theory of relativity must be invoked if the source is moving very rapidly, with a velocity comparable with that of light. If the observation direction is the same as that of the motion, the effect is a *longitudinal Doppler effect.*
An approximation for the relativistic formula for low speeds of the source gives the *linear Doppler* effect. If the propagation direction is not the same as that of the relative motion, and particularly if these two directions are perpendicular, there is a *transverse Doppler effect* or *quadratic Doppler effect.* This has also been observed on canal rays. The Doppler effect is used to measure the speeds of moving objects, e.g. in Doppler radar. It plays an important role in almost all physical and engineering disciplines (Doppler spectral line broadening, Mössbauer effect, etc.).
The optical Doppler effect is of considerable importance in astronomy in the interpretation of spectra (magnitude and direction of the motion of an object, rotation, etc.).
Stark, J., *Ann. Phys.* **21,** 401 (1906)
Stark, J., *Ann. Phys.* **26,** 27 (1908)
Lit.: [9, 10, 40 Vol. I] [76] (additional lit. see Table 39)

Optical Kerr effect L

A very strong linearly polarized light field produces anisotropy in the refractive index in an isotropic medium (usually a liquid). The difference in refractive index parallel and perpendicular to the polarization vector is proportional to the square of the field strength. This is called the optical Kerr effect.
Molecular orientation in a field of light involves times of about 10^{-13} seconds. Then it is possible to make an ultrafast Kerr switch, as required in picosecond emission spectroscopy.
→ nonlinear optical effects
Lit.: [42] (additional lit. see Table 31)

Optical mixing

→ nonlinear optical effects

Optical rectification

→ nonlinear optical effects

Optoacoustic effect (1880) O

(photoacoustic effect)
In 1881, Bell, Röntgen and Tyndall first achieved the conversion of intensity-modulated light to audible sound. This photoacoustic effect is based on the absorption of intensity-modulated light in a gas, which leads to warming or cooling. One thus has a source of heat modulated in the audible range. The effect is used in spectroscopy (optoacoustic methods) and latterly has been used in optical communication systems for the direct conversion of modulated light into sound. The efficiency of the effect is extremely small (10^{-5}). Other applications are given below. Sometimes the effect is called the *thermo-pneumatic effect* and is used in the *Golay detector* for observing thermal radiation.
Applications:
Photoacoustic analysis
Photoacoustic spectroscopy: optical spectra of surface domains of variable thickness;
Photoacoustic thermoanalysis: measurement of thermal diffusivity, phase transitions, and microcalorimetry with high spatial resolution;
Photoacoustic flaw detection: observation of exfoliation and cavities;
Photoacoustic microscopy: observation of microstructures and the effects on local thermal diffusivity.

BELL, Alexander Graham,
Scottish-American physicist and deaf-and-dumb teacher, 3 March 1847 – 1 August 1922, from 1873 Professor in Boston
Bell, A. G., *Am. J. Sci.* **20,** 305 (1880)
Bell, A. G., *Philos. Mag.* **11,** 510 (1881)
Tyndal, J., *Proc. R. Soc.* (London) **31,** 307 (1881)
Röntgen, W. C., *Philos. Mag.* **11,** 308 (1881)
Coufal, H. J. and Lüscher, E., *Phys. unserer Zeit* **9,** 46 (1978)
Lit.: [96, 98, 100, 106]

Optogalvanic effect Pl

(optovoltaic effect)
The voltage across a discharge tube may be influenced by the application of light (irradiation of the discharge space). A load resistance allows either current changes (optogalvanic effect) or voltage changes (optovoltaic effect) to be observed.
The effect has been known for over 50 years and has been used successfully to stabilize the output power of CO_2 lasers.
Scholz, A. L. and Schiffner, G., *Appl. Phys.* **21,** 407 (1980)
Smith, A. L. S. and Moffatt, S., *Opt. Commun.* **30,** 213 (1979)
Lit.: [100, 114]

Optovoltaic effect

→ optogalvanic effect

Osmotic effects

→ colligative effects

Overhauser effect (1953) N

In nuclear magnetic resonance in metals (in particular alkali metals), the resonance amplitude can be increased if a high-frequency field of frequency equal to the Larmor frequency of the conduction-electron spins, is applied simultaneously. The doubly split levels are then not populated in accordance with thermal equilibrium, but have an equal or excess population in the upper energy level. The transition from the higher level intensifies the nuclear signal.
Overhauser, A. W., *Phys. Rev.* **92,** 411 (1953)
Lit.: [75] (additional lit. see Table 38)

Overheating effect

→ undercooling effect

Ovshinsky effects (1968) SCC

In 1967/8, Ovshinsky found that doped amorphous semiconductors show novel effects. He used chalcogenide glasses, which were composed of nonstoichiometric combinations of the chalcogens (S, Se, and Te) with group V elements (As and Sb). Thin films of these chalcogenide glasses exhibited two effects used in electronic components: the *switching* and *memory effects.*
The current-voltage characteristic in the switching effect is similar to that of a discharge lamp. When a threshold voltage is exceeded, the conductivity increases suddenly by more than five powers of 10 on going from the "off" state to the "on" state. When the current is less than a certain critical value, a transition back to the "off" state is observed. On the other hand, if the current in the "on" state is slowly reduced, the

highly conducting state can persist in the current-free state, and the memory effect is said to take place.
In the "on" state of the switching effect, the current flows through a hot filament of fluid material, which on rapid cooling reverts to the amorphous "off" state. On slow current reduction, a crystalline conducting filament persists, and the → memory effect results (Lit.: 1 and 2). The switching process from "off" to "on" has not been fully elucidated. It is assumed that, in addition to thermal effects, there is a fast electronic process conditioned by the amorphous structure of the material (Lit.: 3 and 4).

Jäntsch, O., *Phys. Unserer Zeit* **3,** 10 (1972)
Fritsch, G., *Phys. Unserer Zeit* **5,** 120 (1974)
McGinnes, J., Corry, P., and Proctor, M., *Science* **183,** 853 (1974)
1. Stocker, H. J., *J. Non-Cryst. Solids* **2,** 371 (1970)
2. Guntersdorfer, M., *J. Appl. Phys.* **42,** 2566 (1971)
3. Mott, N. F., *Contemp. Phys.* **10,** 125 (1969)
4. Haberland, D. R., *Frequenz* **27,** 68 (1973)
Lit.: [91] (additional lit. see Table 49)

Packing effect N

Due to the nucleon binding energy, the packing of nucleons together in a nucleus implies a mass reduction in accordance with Einstein's formula, which is called the packing effect.

Lit.: [72, 74] (additional lit. see Table 38)

Pair annihilation

→ pair-production effect

Pair-production effect AQ I

Gamma rays of sufficient energy cause pair production (electron-positron pairs) in the potential field of a nucleus or atom. The threshold for this process is given by Einstein's mass-energy formula. For energies over 1 MeV, pair production plays an increasingly important part in the absorption of gamma rays, in addition to the → Compton effect and → photoelectric effect. The converse effect is *pair annihilation,* the destruction of an electron and positron to give two gamma rays. In both processes, charge, spin, angular momentum, as well as energy and momentum must be conserved.

→ scattering effects, → Compton effect, → photoelectric effect

Lit.: [6c, 16, 17, 79, 90] (additional lit. see Table 20)

Paramagnetic rotation

→ Becquerel effect (1906)

Particle scattering Sc

The scattering of atomic or subatomic particles is one of the principal topics examined in particle or high-energy physics. It is possible to gain an understanding of subatomic forces and particles from the results of scattering experiments with large accelerators. As in the scattering of electromagnetic waves, there are elastic and inelastic scattering processes.
The most important scattering processes are the following:

Bhabha scattering

The scattering of high-energy positrons by electrons is called Bhabha scattering.

BHABHA, H. Jehanger,
Indian physicist, 30 October 1909 – 24 January 1966

Born scattering

If allowance is made for the effects of the electron shells on nuclear scattering in the Born approximation, we speak of Born scattering.

BORN, Max,
German physicist, 12 December 1882 – 5 January 1970, from 1933 Professor in Berlin, Frankfurt, Göttingen and Edinburgh

Delbrück scattering (1933)

Elastic scattering of light quanta at the nuclear potential is known as Delbrück scattering. The photons split into virtual electron-positron pairs in the neighborhood of the nucleus. Following the interaction with the nuclear field, a scattered photon is emitted with pair annihilation. Delbrück scattering cannot be isolated from the other scattering processes, and it always occurs coupled to them.

DELBRÜCK, Max
German-American physicist and biophysicist,
4 September 1906 – 1980

Mott scattering
The scattering of transversely polarized electrons at a Coulomb field is called Mott scattering. The orbital angular momentum of an electron is dependent on whether the electron passes the scattering charge on the right or left. A magnetic moment is associated with the orbital angular momentum, which leads to differences in interaction energy in the two cases. Quantum-mechanical scattering theory gives the correct formula for this process. Mott scattering allows measurement of the electron polarization in β decay of nuclei; longitudinal polarization is found. The scattering of electrons at electrons is called Møller scattering.

MOTT, Sir Neville Francis,
British physicist, 30 September 1905, Professor in Bristol

MØLLER, Christian,
Danish physicist, 22 December 1904, Professor in Copenhagen

Lit.: [6c, 6d, 6e, 7, 9, 13, 22d, 38 Vol. III, IV, V, VI, VIII, X, 45, 49, 50 Vol. II, 2.1, II, 2.2] [50 Vol. 4, part 3, 60, 61 Vol. III, IV, 69, 72 – 79, 90, 93] (additional lit. see Table 47)

Rutherford scattering
(Coulomb scattering)
By this is meant the scattering of an electrically charged particle at the Coulomb field of another charged particle on account of the Coulomb interaction. The scattering of α particles by atoms led Rutherford to formulate an atomic model. Rutherford's scattering formula gives the same result for small α-particle energies both classically and quantum-mechanically. At high energies, quantum mechanics gives a correction term. This correction term allows the nuclear radius to be deducted from scattering experiments.

RUTHERFORD, Ernest,
from 1931 Lord R. of Nelson, British physicist, 30. August 1871 Nelson (New Zealand) – 19 October 1937 Cambridge, Professor in Montreal, Manchester and Cambridge

Lit.: *textbooks*

Table 11. Elastic scattering

	Scattering by	Notes
Mie scattering	particles comparable with or larger than the wavelength: dust, soot, etc.	Characteristic back scattering
Resonant scattering	atoms, molecules	
Nuclear resonant scattering	atomic nuclei	Excitation of shell electrons or nuclear levels
Rayleigh scattering	particles small relative to the wavelength	Inversely proportional to the fourth power of the wavelength;
Tyndall effect	submicroscopic particles in solution	Perpendicular to the incident beam direction the scattered light is linear polarized
Thomson scattering	quasifree electrons	
Nuclear Thomson scattering	atomic nuclei	Photon energies of 0.01 – 10 MeV (photon wavelengths from 10^{-8} cm to 10^{-11} cm)

Table 12. Inelastic scattering

	Scattering by	Relative frequency change
Compton effect	free electrons	$-(10^{-3}$ to $10^{-1})$
Pair production	atomic nuclei	-1
Raman effect	molecules	$\pm(10^{-3}$ to $10^{-2})$
Brillouin effect	crystals and liquids	$\pm(10^{-6}$ to $10^{-5})$

Paschen-Back effect (1912) Mo

(transition effect, conversion effect, magnetooptic conversion)

As a magnetic field is increased, the complicated line splitting in the anomalous → Zeeman effect is transformed (magnetooptic transformation), and at very high field strengths the splitting is reduced to the triplet of the normal Zeeman effect. In the Paschen-Back effect, the increasing magnetic field progressively eliminates the Russell-Saunders coupling between the orbital angular momentum and the electron spin. The orbital momentum and the spin then each perform independent Larmor precessions around the direction of the magnetic field. The spin-orbit interaction causes the lines in the triplet to show fine structure.

The Paschen-Back effect is therefore not only the final state at very high fields but includes the complete transformation of the anomalous Zeeman effect.

The Paschen-Back effect of the hyperfine structure is also called the → Back-Goudsmit effect.

PASCHEN, Friedrich,
German physicist, 22 January 1865 – 25 February 1947, Professor in Tübingen and Berlin 1924 – 1933, President of the State Technical Physics Institute in Berlin.

BACK, Ernst,
German physicist, 21 October 1881 – 20 July 1959, Professor in Hohenheim and Tübingen

Paschen, F. and Back, E., *Ann. Phys.* **39,** 897 (1912)

Paschen, F. and Back, E., *Ann. Phys.* **40,** 960 (1913)

Lit.: [6d, 17, 60 Vol I, Vol. II] (additional lit. see Table 37)

Peak effect S

In certain hard superconductors (for example, V_3Ga), the dependence of the critical current on the external magnetic field shows a peak near the critical magnetic field. The explanation is that normal-conduction regions arise in the neighborhood of the critical magnetic field. These act as trapping centers for flux tubes and cause the critical current to increase.

Lit.: [11] (additional lit. see Table 44)

Peltier effect

→ thermoelectric effects

PEM effect

→ photo-Hall effect

Penning effect (1928)

The Penning effect is that gas atoms or molecules with low electron binding energies can be ionized by excited metastable atoms of higher excitation energy. The ignition voltage for such a gas mixture is lower than the value for the partial gas with the high excitation energy, i.e. with high ignition voltage. For example, the reaction $Ne^* + Ar = Ne + Ar^+ + e$ reduces the ignition voltage for Ne from 750 to 180 V when 0.006 % Ar is added to the Ne.

PENNING, Franz Michel,
12 September 1894 – ?

Penning, F. M., *Z. Phys.* **46,** 335 (1928); *Physica (Haag)* **11,** 183 (1931)

Lit.: [76, 86 Vol. XXII]

Perihelion rotation, secular R

Astronomers have known for centuries that the perihelia in the planetary paths around the Sun shift by small angles over long periods of time (secular times); this observation could not be explained by Newtonian mechanics. Moreover, no modification in mechanics led to an explanation. The effect follows directly from general relativity theory.

The theory provides the values listed in Table 13.

Table 13. Perihelion rotation

Planet	Angle shift
Mercury	43.02″
Venus	8.6″
Earth	3.8″
Mars	1.3″

The astronomical data for Mercury give a value of 42.56″, which Leverrier already measured in 1859. The perihelion advance is one of the best-founded experimental confirmations of the general theory of relativity.

Einstein, A., *Sitz.-Ber. Preuss. Akad. Wiss.* 2. Vol. 831 (1915)

Shapiro, I. I., *Phys. Rev. Lett.* **28,** 1594 (1972)

Lit.: [6c, 10, 59] (additional lit. see Table 43)

Perturbation propagation effect

→ substituent effects

Phonon-drag effect

→ drag effect

Photoacoustic effect

→ optoacoustic effect

Photocapacitative effects PS

If a semiconductor is used as dielectric in a capacitor, there is a change in the capacitance on illumination. The effect is due to the production of charge carriers in the semiconductor bulk (internal photoelectric effect). Similarly, the change in the capacitance of a semiconductor surface film on illumination is a photocapacitative effect.
Lit.: [6d] (additional lit. see Table 42)

Photoconduction

→ photoelectric effects, → internal photoelectric effect

Photodielectric effect PS

At low temperatures a semiconductor (ZnS or $CaWO_4$) shows the photodielectric effect, but this is extremely small because of the small number of polarizable centers. The frequency at which the dielectric constant alters is then dependent only on the temperature, but not on the excitation density.
The effect is very difficult to observe and is masked by the → photocapacitative effects.
Lit.: [6d]

Photoelectric effects P

In general, a photoelectric effect is an interaction of photons with matter. The term photoeffect is therefore applied to the entire electromagnetic spectrum.
Distinctions are drawn between the photoelectric effects observed with low-energy γ rays (x rays), such as ionization of atoms and molecules, with high-energy γ rays (the internal photoelectric effect, the nuclear photoelectric effect, the Auger effect), as well as with very high-energy γ rays (photo production), UV light (external photoelectric effect), and the interaction of light with semiconductors (internal photoelectric effect, photoconduction).

1. External photoelectric effect
Lenard was the first to suggest that the rays produced in the → Hallwachs effect and → Hertz effect were electrons. He found the following simple relationships for the external photoelectric effect (*Maggi effect*), which was later interpreted by Einstein in 1905:

- The electron current is strictly proportional to the absorbed light energy (Hallwachs).
- The effect occurs without delay (10^{-11} s) to its full extent.
- The maximum energy of the released electrons is a linear function of the light frequency.
- The speed of the photoelectrons is independent of the light intensity.
- There is a long-wave limit specific to the material at which the photoelectric effect first occurs.

The name *Lenard-Einstein equation* is given to the relationship between the mechanical energy of the electrons, the energy of the light, and the work function: the energy of motion of the electrons must be equal to the difference between the light energy and the electron work function of the material. By reducing the light energy the long-wave limit is found, at which no electrons are produced, and above which the photoelectric effect takes place. Usually, this is the smaller the more electropositive the metal.
Electron release from metal lattices is also called the *normal photoelectric effect*. It occurs with clean, compact metal surfaces. The sensitivity (photocurrent/incident light) decreases as the wavelength increases.

2. Hallwachs effect (1888)
(light-electric effect)
In 1888 Hallwachs, one of Hertz's students, discovered that a negatively charged metal plate discharges when irradiated with ultraviolet light. There was no effect with a positively charged plate. Hence, the plate must "lose" electrons on irradiation.

HALLWACHS, Wilhelm,
German physicist, 9 July 1859 – 20 June 1922, Professor in Dresden

Hertz, H., *Ann. Phys.* **31**, 983 (1887)
Hallwachs, W., *Ann. Phys.* **33**, 301 (1888)
Hallwachs, W., *Ann. Phys.* **37**, 666 (1889)
Lenard, P., *Ann. Phys.* **8**, 149 (1902)
Einstein, A., *Ann. Phys.* **17**, 132 (1905)

Millikan, R. A., *Phys. Z.* **17,** 217 (1916)
Lit.: [2, 6b, 6c, 6d, 7, 16, 17, 22, 50 Vol. II, 2.2; Vol. 4, part 4] (additional lit. see Table 40)

3. Hertz effect (1887)
In 1887 Hertz discovered that the voltage required to produce a gas discharge was reduced when the cathode was irradiated with ultraviolet light. The effect is derived from the → external photoelectric effect.

4. Internal photoelectric effect
(photo-volume effect, photoconduction)
The external photoelectric effect occurs at the surface of a metal or semiconductor, whereas, in contrast, the photoelectric effect occurring within a sample is known as the photo-volume effect or the internal photoelectric effect. It is of particular importance in semiconductors and is the cause of photoconductivity.
If a homogeneous semiconductor sample with a voltage applied to it is exposed to a homogeneous light beam, the conductivity may alter (*photoconductivity*). Either fresh charge carriers are produced or the mobility of the existing ones is altered.
If the energy of the absorbed light is greater than the band separation, free charge carriers can be produced. If the carrier binding energy corresponds to the width of the forbidden zone, one speaks of lattice absorption, while in other cases one is dealing with defect absorption.
In lattice absorption, mobile charge carriers of both types are produced, whereas only one type is produced in defect absorption.
The photoelectric effects have acquired many applications, examples being photodiodes, photoresistors, and photocells.
The photoelectric effects in semiconductors are considered in the section on this topic.
Lit.: [2, 6d, 17, 41, 46, 49], *textbooks, dictionaries* (additional lit. see Table 40)

5. Selective photoelectric effect (1894)
(spectral selective effect, vector effect)
In thin metal films, the photoelectric effect shows a resonance-type dependence on the incident photon energy, if the light vector has a component perpendicular to a crystal plane (vector effect). With freshly prepared alkali-metal surfaces, a spectral maximum is found in the emission.
Elster, J. and Geitel, H., *Wied. Ann.* **52,** 433 (1894)
Elster, J. and Geitel, H., *Wied. Ann.* **55,** 684 (1895)
Elster, J. and Geitel, H., *Phys. Z.* **11,** 257 (1910)
Lit.: [50, Vol. 4, part 4] (additional lit. see Table 40)

Photoelectric effect at high light intensities L

The normal (external) photoelectric effect applies for radiation intensities of a few Watts per cm^2. At the intensities that can be produced at the focus of a lens when a laser is used as source, gases may be ionized, although the incident energy is much less than the ionization energy. This effect has a complicated dependence on the wavelength, type of gas, and focus diameter. The effect occurs above a critical intensity.
→ nonlinear optical effects
Lit.: [6d] (additional lit. see Table 31)

Photogalvanic effect

→Dember effect

Photogalvanomagnetic effect

→ photo-Hall effect

Photo-Hall effect (1934) PS

(Kikoin, Noskov, Frenkel)
(PEM effect, photoelectromagnetic effect, photogalvanomagnetic effect)
If a magnetic field perpendicular to the beam direction is applied in a Dember apparatus, the diffusing electrons and holes within the material are deflected in opposite directions. A potential difference is thus produced perpendicular to the magnetic field and to the beam. A current can then flow in an external circuit without a driving voltage (*PEM effect*).
The *photo-Hall effect* is the change in the Hall voltage on illuminating a semiconductor. It is thus a demonstration of the Hall effect on the photocurrent.
Kikoin, I. K. and Noskov, M. M., *Phys. Z. Sowjet.* **5,** 586 (1934)
Frenkel, J., *Phys. Z. Sowjet.* **5,** 597 (1934); **8,** 185 (1935)
Lit.: [2, 6d, 87 Vol. XIX, XX] (additional lit. see Table 42)

Photoluminescence
→ luminescence effects

Photon-drag effect
→ drag effect

Photorefractive effects O
Many materials show distinct changes in properties on exposure to intense light beams. Phenomena in which the refractive index alters locally by a few percent in the presence of light are termed photorefractive effects. Certain crystals also exhibit the property of remembering this change. In particular, ferroelectric crystals have a photorefractive memory. Lithium niobate, for example, may be lightly doped with iron, manganese, or copper (one foreign atom per few thousand cells). It is assumed that both the divalent and trivalent forms of iron are present in the crystal. The photons release an electron from the divalent iron, which enters the conduction band and moves in response to an external field, or diffusion or the anisotropy in the crystal. After a certain time, such an electron is taken up again by the trivalent iron. The drifting electrons produce internal electric fields, and therefore local changes in the refractive index via the → electrooptic effect.
The changes disappear after a time ranging from a few minutes to years. Heat treatment, however, enables the changes to be stored permanently.
Lasers enable → holograms to be stored in these crystals. At present it is possible to produce $10^{12}-10^{13}$ bit cm $^{-3}$, i.e., several orders of magnitude larger than in current semiconductor memories.
Examples of such photorefractive crystals are: $LiNbO_3$: Fe (lithium niobate) and $LiTaO_3$: Fe (lithium tantalate)
Schmitt, H. J., *Phys. unserer Zeit* **12,** 3 (1981)
Lit.: [95, 100]

Photovoltaic effect
→ barrier-layer photoelectric effect

Photovolume effect
→ photoelectric effect

Piezocaloric effect ET
The piezocaloric effect describes the entropy produced when a mechanical stress occurs in a crystal. The converse is known as thermal expansion.
The effect has no direct applications, and it is generally an interfering effect.
Mageri, R. and Riedel, C., *Elektrie* **29,** 308 (1975)
Lit.: [95, 100, 102, 110]

Piezoelectric effect (1880) ET
(piezoeffect)
Piezoelectric effects occur in any crystal lacking a center of symmetry. Such a crystal has a polar axis, i.e., a direction that is physically distinct from the opposite direction. The effects derive from the electrical polarization in a preferred direction in response to mechanical deformation (*direct piezoelectric effect,* the Curie brothers 1880). The converse may also be observed: when an electrical potential is applied, the crystal changes in shape (*inverse piezoelectric effect,* Lippmann 1881). The effect has been demonstrated on many different crystals (Seignette salt, willemite, quartz, zinc blende, tartaric acid, sodium chlorate, tourmaline, and sucrose). At present, the following substances are of considerable technical importance: GaAs, ZnO, CdS, $LiNbO_3$, $LiTaO_3$, and quartz.
Distinctions are drawn between the *longitudinal, transverse,* and *shearing effects,* in accordance with the direction of the force component and that of the polarization of the material. This applies to the piezoelectric effect and its converse.
The piezoelectric effect has found many applications (Table 14). In addition to classical materials, ferroelectric ceramics (lead zirconate-titanate) are of the greatest technical significance. The piezoelectric effect is closely related to the pyroelectric effect and to electrostriction.
→ acoustoelectric effect

CURIE, Pierre,
French physicist, 15 May 1859 – 19 April 1906, from 1904 Professor at the Sorbonne

CURIE, Jacques Paul,
29 October 1855 Paris – 29 February 1941 Montpellier

Curie, J. and Curie, P., *C. R.* **91,** 294, 3838 (1880)
Curie, J. and Curie, P., *C. R.* **92,** 186 (1881)
Curie, J. and Curie, P., *C. R.* **93,** 204 (1881)
Lippman, G., *J. Phys. Paris* 61, **10,** 391 (1881)

Pockels, F., *IV. Jahrb. Mineral., Suppl.* 7, 224 (1890)
Guntersdorfer, M., *Phys. unserer Zeit* **7**, 48 (1976)
Lit.: [6b, 18, 22b, 27, 34, 37, 38 Vol. VIII, 69, 72 – 78, 90 – 92] (additional lit. see Table 28)

Piezojunction effect SC

A p-n junction shows a change in current-voltage characteristic in response to mechanical stress. This piezojunction effect can be used in the detection of acoustic signals or mechanical stresses.
Lit.: [92] (additional lit. see Table 48)

Table 14. Piezoelectric effect: applications

Direct piezoelectric effect	Inverse piezoelectric effect	Combination of the direct and inverse piezoelectric effects
Buttons	Ultrasonic sources	Distance measurement
Gas ignition	Cleaning	Electromedicine
Sound heads	Material processing	Echo sounding
Microphone	Material testing	Surface wave techniques
Force and acceleration measurement	Remote control	Filters
	Adjustment of video heads	Telephony entertainment electronics
	Laser mirrors	Remote control
	Relays	High-voltage transformers
	Oscillators	Isolating transformers
	Shock excitations	Conduction delay
	Liquid spraying	
	Color printing	Buffer memories
	Telephone ear pieces	Signal compression
	Audio filters	
		Graphics display
		Presence detector
		Concentration measurement on vapors

Piezomagnetic effect Mm

Piezomagnetism means the various reversible → magnetomechanical effects. In particular, it means that an alternating magnetic field of small amplitude acting on a magnetized sample produces a change in shape proportional (linearly) to the magnetic field, and also the converse: namely, that a mechanical oscillation in a magnetized sample produces a change in the magnetization proportional to the extension.
Lit.: [92] (additional lit. see Table 36)

Piezoresistance effect ET

The change in electrical resistance of a metal or semiconductor produced by a mechanical stress is known as the piezoresistance effect. The cause lies in the band structure in a semiconductor or in the Fermi surface in a metal, which is altered in shape by the mechanical stress and so affects the conductivity.
Lit.: [92] (additional lit. see Table 28)

Pinch effect Pl

A plasma carrying a large current tends to contract in a strong magnetic field on account of its own magnetic field. Two arrangements may be distinguished:
axial magnetic field and azimuthal current (Θ pinch)
radial magnetic field and axial current (Z pinch)
The directions of the magnetic field and the current are selected so that the resultant force compresses the plasma.
Lit.: [6c] (additional lit. see Table 41)

Plasma effects Pl

The name plasma is given to a state of matter characterized by ionization and quasineutrality; with appropriate energy input, the electrons are separated from the atoms, and ionized matter results. A given volume contains neutral atoms, ions, and electrons in certain numbers. If there is

no space charge in the region, one speaks of *quasineutrality*.
Energy can be supplied in various ways: current flowing through a gas, HF irradiation, laser beams, rapid gas compression, chemical reactions, or nuclear reactions.
The properties of plasmas are controlled by the simultaneous occurrence of charge carriers of both types. The following are affected: the propagation of electromagnetic and electrohydrodynamic waves, transport properties such as heat conduction, viscosity, diffusion, and electrical conductivity, and the emission of electromagnetic radiation in the complete frequency range down to X rays.
Plasma physics links up various fields in physics:

Hydrodynamics		Atomic physics
Electrodynamics	with	Nuclear physics
Thermodynamics		Statistical mechanics

Thus, many effects occur in other contexts.
→ pinch effect, → runaway effect
Lit.: [100, 114]

p-n photoelectric effect

→ barrier-layer photoelectric effect

Pockels effect (1893) Eo

(linear electrooptic effect)
The Pockels effect is a linear → electrooptic effect, which is observed in piezoelectric crystals such as ammonium dihydrogen phosphate or potassium dihydrogen phosphate, which do not have centers of symmetry. The electric field acts longitudinally. The resulting birefringence is directly proportional to the applied field. The sign of the refractive index change also reverses when the field is reversed. The Pockels effect is used in modulating and switching giant-pulse lasers.

POCKELS, Friedrich Carl,
German physicist, 1865 – 1913
Pockels, F., *Gött. Abh.* **39,** (1893)
Lit.: [19, 48, 57, 76, 77] (additional lit. see Table 27)

Poisson effect F

(pad effect)
A spinning projectile tends to compress the air on one side but to expand it on the other. Because of the increased friction, the projectile begins to roll on the denser air. With right-handed spin, it is deflected to the right on the air cushion, and vice versa.

POISSON, Siméon Denis,
French physicist and mathematician, 21 June 1781 – 25 April 1840, from 1806 Professor in Paris
Lit.: [75] (additional lit. see Table 29)

Polarization effect in mass spectroscopy

If cathode rays or canal rays strike metal parts under vacuum, they generally produce insulating films at these points, which become charged by the beam and can then cause beam deflections. This gave rise to considerable experimental difficulties, especially in early mass spectrographic research work.
Lit.: [76]

Pole effect Eo

A slight shift in spectral line wavelength near the poles of a light source is observed relative to the lines emitted in the middle of the source. The effect is due to the high field gradients near the poles and the strong electric fields produced by ions there.
Takamine, T., *Astrophys. J.* **50,** 23 (1919)
Lit.: [50 Vol. II, 2.2, 76] (additional lit. see Table 27)
→ Poisson effect

Pomeranchuk effect LT

The helium isotope ^3He has the property that the liquid phase is more ordered than the solid phase below O.3 K. Consequently, the entropy of the liquid is less than that of solid ^3He. On adiabatic compression of ^3He, the temperature must therefore fall until all the ^3He has solidified (the Pomeranchuk effect). This method has been successfully used to attain temperatures in the range from 1 mK to 30 mK.
→ Onnes effects, → superconduction, → electrocaloric effect, → magnetocaloric effect

POMERANCHUK, Isaak Yakovlevich,
Russian physicist, 1913
Lit.: [6d] (additional lit. see Table 33)

Poole-Frenkel effect (1938) SC

The Poole-Frenkel effect is concerned with the increase in electrical conductivity in insulators and semiconducting materials as the result of field effects. The effect is due to a mechanism similar to that of the Schottky effect, in which there is a reduction in the effective electron work

function at a metal surface because of an applied electric field.
Frenkel, S., *Phys. Rev.* **54,** (1938); *Techn. Phys. of the USSR* **5,** 685 (1938)
Ieda, M. et al., *J. Appl. Phys.* **42,** 3737 (1971)
Lit.: [100, 104, 110, 114]

Portevin-Le Chatelier effect (1923) So

Foreign atoms in a material have a distinct effect on the shape and position of the hardening curve (shear stress or stress as a function of strain) and also on the elastic limit of the material. The occurrence of stepouts on the smooth curve is called the Portevin-le Chatelier effect, whereas the occurrence of a sharp elastic limit is called the *elastic-limit effect*. In the Portevin-le Chatelier effect, diffusing point defects interfere with a dislocation that has already been checked in its motion by other obstacles. The interplay of blocking and release leads to the steps in the stress-strain diagram. The Portevin-le Chatelier effect is the mechanical analogue of the → Barkhausen effect.

LE CHATELIER, Henri Louis,
French metallurgist and chemist, 1850 – 1936
PORTEVIN, Albert Marcel Germain,
French metallurgist and chemist, 1 October 1880 – 1962
Portevin, A. and le Chatelier, F., *C. R. Acad. Sci. Paris* **176,** 1, 507 (1923)
Lit.: [37, 52a, 72] (additional lit. see Table 50)

Poynting effect So

Torsion of a very long cylindrical rod produces an effect on the length (second-order effect).

POYNTING, John Henry,
British physicist, 9 September 1852 – 30 March 1914, from 1880 Professor in Birmingham
Lit.: [74] (additional lit. see Table 50)

Primakoff effect AQ II

The decay of a neutral meson into two gamma rays represents the time reversal of the photoproduction of this meson at a nucleus, with the second photon represented by the Coulomb field of the nucleus. The cross section for this production process is related to the meson lifetime, and lifetimes of 10^{-16} S and less, as are characteristic of electromagnetic decay, can be measured in this way.

PRIMAKOFF, Henry,
American physicist, born 12 February 1914 in Odessa
Primakoff, H., *Phys. Rev.* **81,** 899 (1951)
Lit.: [100, 114, 118a]

Printout effect SC

A silver halide crystal becomes dark on illumination because of colloidally deposited silver (printout effect). This effect can be used, for example, to measure the electron drift velocity. The crystal is exposed to short-wave light, which releases electrons only in a thin surface layer. These are drifted into the crystal by an applied electric field. They are made visible by the printout effect. The drift velocity can be deduced from the voltage application time and the path travelled.
Lit.: [22d] (additional lit. see Table 48)

Proximity effect S

(coupling effect)
If a normally conducting layer borders on a superconductor, it contains Cooper pairs in small numbers, since the order parameters must change continuously in the boundary layer.
The same name is given to all phenomena that occur when the adjoining layers are ferromagnetic and metallically conducting, or ferromagnetic and insulating, or ferroelectric and insulating.
The transition temperature of the superconductor is also altered by the proximity effect.
Lit.: [75] (additional lit. see Table 44)

Pyroelectric effect ET

(pyro effect)
A ferroelectric crystal shows surface electric charges of both signs in response to rapid temperature change. This indicates that the spontaneous polarization is exceedingly temperature-dependent. Any ferroelectric material is simultaneously pyroelectric and piezoelectric. Examples are $LiTaO_3$, triglycine sulfate, ferroelectric ceramics, and polyvinylidene fluoride films. The converse does not apply: quartz (piezoelectric) is neither pyroelectric nor ferroelectric.
This effect interferes in the application of the piezoelectric effect. It is used to detect and measure infrared radiation in the range from 1 μm up to 40 μm, and in that case the accompanying piezoelectric effect is an interfering effect. It is also used in infrared image converters. The converse

of the pyroelectric effect is the *electrocaloric effect.*
If an electric field is applied to tourmaline, for example, there is a temperature change in the crystal.
Lit.: [6b, 50 Vol. 4, part 1] [69, 74, 75] (additional lit. see Table 28)

Quadratic Doppler effect

→optical Doppler effect

Radiofrequency effect

→ size effect

Radiometer effect (1825) TK

Surfaces at different temperatures in a low vacuum show a pressure difference because of differences in momentum transfer to gases colliding with them. The effect can be used to detect heat radiation. A well-known application is its use in the Crookes radiometer.
The effect was discovered in 1825 by Fresnel, who observed that a body suspended in a vacuum is deflected backwards on exposure to light. Crookes examined the effect more precisely and constructed the radiometer known after him. In the Crookes radiometer, a vacuum of $10^{-5}-10^{-6}$ bar is used with four glass plates, each of which is blackened on one side and is suspended so that it can rotate; the blackened side is deflected backwards on irradiation.
Fresnel, A., *Ann. Chim. Phys.* 29, 57, 107 (1825)
Crookes, W., *Philos. Trans. R. Soc. London* **164,** 501 (1874)
Crookes, W., *Philos. Trans. R. Soc. London* **170,** 132 (1880)
Crookes, W., *Philos. Trans. R. Soc. London* **166,** 380 (1876)
Hettner, G., *Ergeb. Exakten Naturwiss.* **7,** 209 (1928)
Lit.: [50 Vol. III, 2] (additional lit. see Table 54)

Rainbow effect (1959) Sc

In the scattering of particles at a potential that contains attractive and repulsive parts, scattering results that is completely analogous to the scattering of light at liquid droplets; there is a pronounced maximum in the scattering intensity at a certain angle (primary rainbow). On one side of the peak, there is a sharp fall in the intensity (dark side of a rainbow), while the intensity on the other side oscillates (secondary rainbows).
Rainbow scattering occurs when the particle wavelength is small relative to the width of the potential, and the potential changes only slightly within a wavelength.
Scattering measurements allow one to determine the potential parameters, e.g., for a Lennard-Jones potential.
→ scattering effects, particle scattering
Ford, K.W. and Wheeler, J. A., *Ann. Phys. (N. Y.)* **7,** 259, 287 (1959)
Hundhausen, E. and Pauly, H. *Zeitschr. f. Physik* **187,** 305 (1965)
Lit.: [98, 100, 109, 114]

Raman effect

→Raman-Smekal effect

Raman-induced Kerr effect L

As in the → AC Kerr effect, Raman-induced birefringence for the observation beam occurs when the two optical frequencies (observation and pumping beams) differ by a characteristic frequency of the medium (vibration or rotation).
→ Kerr effect, → Raman effect, → nonlinear optical effects
Hellwarth, R. W., *Progr. Quant. Electron.* **5,** 1 (1977)
Lit.: [98, 106, 119]

Raman-Smekal effect (1928) Sc

(Raman effect)
When monochromatic light is scattered by atoms or molecules, characteristic frequencies which need not be in resonance with the energy levels are observed in the scattered radiation. It is found that the frequency difference between the primary light and the scattered light is independent of the primary frequency. Also, the difference corresponds to the infrared vibrations of the scattering medium. The frequencies on the long-wave side of the primary line are known as *Stokes lines,* while those on the short-wave side are known as *anti-Stokes lines.* The Raman effect is concerned with the inelastic scattering of photons by matter.
The effect was predicted in 1923 by Smekal and was observed in 1928 by Raman, and almost simultaneously by Landsberg and Mandelstam. Electromagnetic waves produce a polarization of the material dependent on the displacement of the electrons in the atoms. Expansion of the po-

larization as a function of displacement in a series provides various effects. In the first-order Raman effect, only one photon is involved, whereas two photons are required in the second-order process.
Raman spectroscopy of atoms and molecules provides an understanding of the binding forces and of complicated molecular spectra. The Raman effect in rotation and vibration bands has been called the *rotational* or *vibrational Raman effect*.
In the normal Raman effect, there is no angular dependence. When intense laser beams are used, there is forward scattering, in which Stokes and anti-Stokes lines are observable with high intensity. This is called *induced Raman scattering*. The effect has been used to detect pollutants in the atmosphere.
Resonant Raman effect:
On excitation near an absorption band, Raman lines are found whose intensity increases the closer one comes comes to the absorption line. This is called the resonant Raman effect.
→ Brillouin scattering

RAMAN, Sir Chandrasekhara Venkata
Indian physicist, 7 November 1888 – 21 November 1970, Professor in Calcutta and Bangalore

SMEKAL, Adolph Gustav Stephan,
Austrian physicist, 12 September 1895 – 7 March 1959, Professor in Vienna, Halle, Darmstadt and Graz

LANDSBERG, Grigorii Samuilovich,
12 January 1890 – 2 February 1957

MANDELSTAM, Leonid Isaakovich,
Russian physicist, 4 May 1879 – 27 November 1944

Raman, C. V., *Indian Phys.* **2,** 387 (1928)
Raman, C. V. and Krishnan, K. S., *Nature* **21,** 501 (1928)
Smekal, A., *Naturwissenschaften* **11,** 873 (1923)
Kramers, H. A. and Heisenberg, W., *Z. Phys.* **31,** 81 (1925)
Lit.: [6c, 9, 27, 50 Vol. II, 2.2.) (additional lit. see Table 46)

Ramsauer effect (1920) AQ II

Atomic collision cross sections show anomalies for slow electrons: below 1 eV the cross section as a function of electron speed assumes very small values. The effect is explained as quantum-mechanical scattering of the electrons at the shell electrons, as first proposed by Elsasser; a similar effect occurs in neutron scattering by atoms. The de Broglie wavelength of a slow electron is of the same order as the size of an atom. When the wavelength is of the same order as the size of the obstacle, diffraction and scattering effects become particularly prominent.

RAMSAUER, Carl Wilhelm,
German physicist, 6 February 1879 – 24 December 1955, Professor in Heidelberg, Danzig and Berlin

ELSASSER, Walter,
German-American physicist, 20 March 1904

Ramsauer, C. *Phys. Z.* **21,** 576 (1920)
Ramsauer, C., *Phys. Z.* **22,** 613 (1921)
Ramsauer, C. W., *Phys. Z.* **28,** 7 (1927)
Ramsauer, C. W., *Ann. Phys.* **72,** 8 (1923)
Elsasser, W., *Naturwissenschaften* **13,** 711 (1925)
Lit.: [50 Vol. 4, 3] [75] (additional lit. see Table 21)

Ranque effect F

If a tube having a lumen of a few millimetres is supplied with air having a pressure of a few atmospheres via a tangential nozzle, a coaxial gas vortex occurs, whose outer part is warm and whose core is cold. On entry into the tube, the air cools (→ Joule-Thomson effect). The gas vortex has a lower speed at the edge of the tube than at the axis. The speeds tend to equalize along the tube. This produces warming in the outer zone and cooling at the core. Temperature differences of a few 100 K may be observed.
An important point is that the gas flow is *compressible,* i.e., the speed is close to or above the speed of sound.
Applications: the *Ranque-Helsch tube* allows the separation of a gas jet into hot and cold parts.

Hilsch, R., *Z. Naturforsch.* **1,** 208 (1946)
Lit.: [76] (additional lit. see Table 29)

Rayleigh scattering (1900) Sc

In contrast to Mie scattering, in Rayleigh scattering the radius of the scattering dielectric particle is very small relative to the wavelength. Rayleigh scattering is inversely proportional to the fourth power of the wavelength and explains the blue color of the sky. On observation perpendicular to the beam direction, it is found that the scat-

tered light is linearly polarized. If the particle diameter is larger, Rayleigh scattering is replaced by Mie scattering (→ Mie effect). Rayleigh scattering is closely related to the → Tyndall effect.
→ scattering effects

RAYLEIGH, Lord, John William Strutt, 3rd Baron R.,
British physicist, 12 November 1842 – 30 June 1919, 1879 – 1884 Professor in Cambridge, from 1887 in London
Lit.: [6c, 38 Vol. II, Vol. VIII, 57, 74, 76, 90, 93] (additional lit. see Table 45)

Rayleigh-Taylor effect Sy

(instability)
If a layer of heavy liquid is placed above a lighter one in a gravitational field, even a very slight perturbation causes the liquid layers to interchange, and the labile equilibrium is disturbed. A similar effect occurs with a plasma kept floating by a magnetic field.
Cap, F.: *Introduction to Plasma Physics,* Vol. II [in German]. Vieweg, Brunswick 1972
Lit.: [102a, 103a and b, 112a, 116a]

Recession effect

→ red shift

Rectifier effect (1875) SCC

(barrier effect, barrier-layer effect)
In 1875, F. Braun discovered that the resistance of a metal-semiconductor combination is dependent on the current direction, e.g. a point of bronze, gold, or steel on lead sulfide PbS (MoS_2 or FeS). Such a device is used as a rectifier or as a detector and for the demodulation or rectification of high-frequency alternating current. In 1939, Schottky provided an explanation for the barrier effect in a metal-semiconductor combination, and for that reason this effect is often called the *Schottky effect*.

BRAUN, Karl Ferdinand,
German physicist, 6 June 1850 – 20 April 1918, Professor in Marburg, Strasbourg, and Tübingen
Braun, F., *Pogg. (Wied) Ann. Phys.* **154,** 35 (1875)
Lit.: [6b, 22b, 77] (additional lit. see Table 49)

Red shift As

The spectra of many cosmic objects exhibit red shifts, which are normally explained in terms of the → optical Doppler effect (object receding from the Earth).
When comparing the brightness of two objects, the following factor should be taken into consideration: if one of the two objects is at rest, the moving one will have the lower brightness if it is receding from us, since we receive fewer light quanta per unit time. This is called the *number* or *dilution effect.* On account of the Doppler effect, the frequency of the light is lower, and therefore also the quantum energy. This is called the *energy effect.*
Another red shift occurs when light quanta pass through the gravitational field of a massive star (*relativistic red shift*).
→ gravitational frequency shift
Lit.: [76] (additional lit. see Table 22)

Rehbinder effect (1972) So

Surface-active films can reduce the hardness and ductility of a material.
Rehbinder, P.A. and Shuchkin, E. D., *Prog. Surf. Sci.* **3,** 97 (1972)
Lit.: [52a] (additional lit. see Table 50)

Relaxation effects G

(aftereffects, hysteresis effects)
Relaxation refers to various aftereffects: a body responds to a change in an external force not instantaneously but with a certain delay (*relaxation time*). It approaches a new state of internal equilibrium only asymptotically, and the changes in the relevant parameters usually follow exponential laws.
A material shows inelastic behavior when the stresses and strains are time-dependent. Also, hysteresis is a characteristic response observed when a material is exposed to alternating stresses. Relaxation plays an important part in solid-state effects as well as in many electrical, magnetic, optical, laser, atomic, and nuclear effects.
→ Barkhausen effect, → Portevin-le Chatelier effect, → magnetic aftereffects → Memory effect
Lit.: [37, 76] (additional lit. see Table 19)

Renninger effect (1937) SS

(indirect excitation)

When a crystal is analyzed with X rays or neutrons, a strong diffracted beam can act as a primary beam. This is known as the Renninger effect. It can sometimes lead to the occurrence of normally prohibited reflections. The effect occurs particularly in the analysis of thick crystals. It can be neglected for thin crystals. On the other hand, it always occurs in electron diffraction.

In order to ascertain whether indirect excitation is occurring, the crystal is rotated so that the Bragg condition is always obeyed. If the diffracted-beam intensity is dependent on the angle of rotation, the Renninger effect is present.

RENNINGER, Moritz Karl,
German physicist, 8 June 1905
Renninger, M., *Z. Phys.* **106**, 141 (1937)
Lit.: [25] (additional lit. see Table 51)

Resistance effect

→ galvanometic effects

Resistance-pressure-tension effect EM

The resistivity of a metallic conductor alters as the hydrostatic pressure is increased. Most metals show a decrease in resistivity as the pressure increases. The compression brings the metal atoms closer together, and the binding forces at the equilibrium sites increase. The concept that the resistivity at elevated temperatures is reduced by the thermal vibration of the metal atoms provides a qualitative interpretation of the decrease in resistance as the pressure increases.
Lit.: [76] (additional lit. see Table 26)

Resonant absorption

→ Mössbauer effect

Resonant Raman effect

→ Raman-Smekal effect

Resonant scattering

→ scattering effects

Richardson effect (1903, 1911) EM

(thermionic effect, the Edison effect 1879)

The Edison effect is the American name for the thermionic or Richardson effect, which describes the emission of electrons from heated metals. The only electrons that can leave the metal are those with velocity components perpendicular to the metal surface which are so large that the kinetic energy of the electrons is greater than the work function.

Richardson derived the law for the saturation current in thermionic emission. The saturation current is proportional to the square of the absolute temperature and to an exponential factor: $i_s = AT^2\exp(-W/kT)$, where A is a universal constant and W is the work function (Richardson's law).

In deriving Richardson's law, a Fermi distribution is assumed for the electron velocity distribution within the metal; a Maxwell distribution gives $T^{1/2}$ in addition to the Boltzmann factor. It is difficult to decide between $T^{1/2}$ and T^2 by experiment, since the Boltzmann factor predominates. The electron velocities outside the metal correspond to a Maxwell distribution.

The value for the work function given by the Richardson effect agrees with that obtained from the → photoelectric effect. The effect has worldwide application in the production of light electric bulbs and in the production of free electrons in a vacuum (radio tubes, transmitter tubes, etc.)

→ cooling effect, → Edison effect

RICHARDSON, Sir Owen Williams,
British physicist, 26 April 1879 – 15 February 1959, from 1906 Professor in Princeton (N.J.), from 1914 in London

EDISON, Thomas Alva,
American inventor and industrialist, 11 February 1847 – 8 October 1931

Richardson, O. W., *Philos. Trans. R. Soc. London Ser. A* **101**, 497 (1903)
Richardson, O. W., *Philos. Mag.* **23**, 594 (1912)
Richardson, O. W., *Philos. Mag.* **27**, 476 (1914)
Lit.: [16, 50 Vol. 4, part 4, 84 Vol. IV, 87 Vol. XXI] (additional lit. see Table 26)

Richardson-Barnett effect

→ Barnett effect

Richter aftereffect

→ magnetic aftereffects

Riehl effect Lu

If a phosphor of the recombination type (for example, ZnS or CdS) is exposed to increasing radiation intensities (for example, in the UV range), the luminescence yield increases; simultaneously, temperature quenching is shifted to higher temperatures. The reason for this behavior is that these phosphors exhibit emissive recombination which increases as the square of the number of charge carriers produced by the excitation (second-order reaction), whereas radiationless recombination arising from capture centers increases only linearly (first-order reaction).

RIEHL, Nikolaus,
German physicist, 24 May 1901, St. Petersburg, in Munich since 1957
Lit.: [78]

Righi-Leduc effect

→ thermomagnetic effects

Righi-Leduc effect, second

→ thermomagnetic effects,
→ Maggi-Righi effect

Rim effect

→ gravitational frequency shift

Röntgen current

→ Röntgen-Eichenwald experiment

Röntgen-Eichenwald experiment (1888) EM

In this experiment, Röntgen observed in 1888 that polarization charges produce the same magnetic effect as the corresponding conduction currents. He rotated a dielectric in a charged capacitor. Lorentz called the current produced by the motion of the polarization charges a *Röntgen current*. This is to be distinguished from the Rowland current (→ Rowland effect). To compensate for the Röntgen current, Eichenwald rotated the entire apparatus.
The Röntgen and Rowland currents, like the associated effects, follow from relativistic electrodynamics.

RÖNTGEN, Wilhelm Conrad,
German physicist, 27 March 1845 – 10 February 1923, from 1875 Professor in Hohenheim, 1876 in Strasbourg, 1878 in Giessen, 1888 in Würzburg, and 1899 in Munich

EICHENWALD, Alexander,
Russian physicist, 23 December 1863 St. Petersburg – 1944 Milan
Röntgen, W. C., *Sitz.-Ber. Preuss. Akad. der Wiss,* Vol. 1, 195 (1885)
Röntgen, W. C., *Ann. Phys.* **35,** 264 (1888)
Eichenwald, A., *Ann. Phys.* **11,** (1), 241 (1903)
Lit.: [61 Vol. III] (additional lit. see Table 26)

Rotation effect, magnetic Mo

In a very weak magnetic field, the line splitting in the → Zeeman effect is substantially smaller than the Doppler broadening. Therefore, the splitting cannot be observed. It can, however, be detected by observation of the polarization in fluorescence. The fluorescent light is observed perpendicular to the weak magnetic field and to the beam propagation direction, where the polarization vector of the incident light is also perpendicular to the observation direction. The plane of polarization of the fluorescent light is rotated.
This effect can be explained by the motion of a classical oscillator in a magnetic field which performs a rosette movement when the field and the oscillation direction are mutually perpendicular.
→ Hanle effect
Lit.: [50 Vol. II, 2.2] [77] (additional lit. see Table 37)

Rotation magnetism

→ Arago's experiment

Rowland current

→ Rowland effect

Rowland effect (1878) EM

If a charged circular condenser plate rotates, convection currents arise (Rowland current), which Rowland detected with a magnetic needle. The effect of the convection currents is identical with that of conduction currents which transport the same amount of charge in the same time.
→ Arago's experiment
→ Röntgen-Eichenwald experiment

ROWLAND, Henry Augustus,
American physicist, 27 November 1848 – 16 April 1901 from 1875 Professor in Baltimore
Rowland, H. A., *Philos. Mag.* 445 (1889)
Lit.: [40 Vol. I, 50 Vol. 4, 1; 55, 61 Vol. III, 76] (additional lit. see Table 26)

Runaway effect Pl

When a current flows in a plasma, the plasma is heated by the Joule effect. The conductivity of a plasma is proportional to $T^{3/2}$. This applies only when the electrons in the high-energy part of the Maxwell distribution have no effect on the conductivity. These electrons are called runaway electrons.

Lit.: [84] (additional lit. see Table 41)

Rutherford scattering

→ particle scattering

Sagnac effect (1913) O

If two coherent light beams pass round a surface in opposite directions, interference fringes occur where they overlap. If the entire apparatus is rotated about an axis perpendicular to the surface, the interference fringes shift. The shift enables conclusions to be drawn on the state of rotation of the apparatus with respect to an inertial system. Attempts are being made with lasers to use the effect to replace gyroscopes.

→ Fizeau effect

SAGNAC, Georges Marc Marie,
French physicist, 1869 – 26 February 1928

Post, E. J., *Rev. Mod. Phys.* (2), **39,** 475 (1967)
Sagnac, G., *C. R.* **157,** 708, 1410 (1913)
Sagnac, G., *J. Phys. Radium* (5), **4,** 177 (1914)
Schiffner, G., *Siemens Forschungs- und Entwicklungsberichte* Vol. 9, no. 1, 16 (1980)

Lit.: [40 Vol. I, 33] (additional lit. see Table 39)

Salting-in effect

→ salting-out effect

Salting-out effect TK

The solubility of a nonelectrolyte is reduced by the addition of electrolytes, since the latter form hydration shells, and thereby reduce the amount of water available. The converse is the *salting-in effect.*

Lit.: [35] (additional lit. see Table 54)

Saturation magnetostriction

→ magnetostriction

Scattering effects G, Sc

The diffraction of light (electromagnetic waves) at small particles whose diameters are comparable with the wavelength is called scattering. The incident wave is deflected from its original direction, and the polarization of the scattered radiation may be altered. The angular dependence of the scattered radiation relative to the incident wave is characteristic of the scattering particles.

A distinction is made between *single* and *multiple scattering,* in accordance with the number of scattering partners involved. If the scattering angle is less than or equal to 90°, one speaks of *forward scattering,* while at larger angles we have *backward scattering.* A distinction is also drawn between *elastic* and *inelastic scattering.* In elastic scattering, the frequency is unaffected, and the phases of the incident and scattered beam are definitely connected (elastic scattering therefore is a coherent effect).

The scattering of light (electromagnetic waves) occurs solely at the elastically bound electrons in the atoms or molecules. The characteristic scattering parameters are the photon energy in relation to the electron rest energy and the resonance frequency of the particles. At high light frequencies, the electrons can be treated as virtually free.

Inelastic scattering involves a frequency shift in the scattered light. The photons give up part of their energy or gain additional energy in the scattering process. Consequently, light with longer, resp. shorter wavelength is observed in addition to the original light. At low incident light intensities inelastic scattering is an incoherent effect.

→ Compton effect, → Raman effect

Lit.: [19 and 45]

Scattering of particles

→ Particle scattering

Schadt-Helfrich effect

→ liquid-crystal effect

Schoch effect (1950) Me

This is an acoustic effect corresponding to the → Goos-Hähnchen effect in optics. When acoustic waves were totally reflected from a water-aluminum or xylene-aluminum system, Schoch observed a beam shift with ultrasonic waves. At 16 MHz there was a shift of 2.7 mm, and at 5.5 MHz of almost 12 mm, with an ultrasonic beam diameter of 12 mm.

SCHOCH, Arnold
1911 – 23 July 1967
Schoch, A., *Ergeb. Exakten Naturwiss* **23**, 127 (1950)
Lotsch, H. K. V., *Optik* **32**, 299 (1971)

Schottky effect (1935) SS

Schottky effect is the name sometimes given to positive and negative ions leaving their positions in the crystal lattice and depositing on the crystal surface, while the resulting vacancies are called *Schottky defects.* The number of defects increases with temperature, but the ideal lattice structure is not attained at low temperatures because the defects are frozen in.

SCHOTTKY, Walter,
German physicist, 23 July 1886 – 4 March 1976, 1923 Professor in Rostock, from 1927 with Siemens and Halske
Lit.: [50 Vol. 4, part 4] [75] (additional lit. see Table 51)

Schottky effect SC

Internal field emission can occur in a semiconductor when very high field strengths occur at boundaries such as sharp p-n junctions.
→ avalanche effect, → tunnel effect, → Zener effect, → shot effect
Lit.: [94, 100, 104, 110, 114]

Schottky effect SCC

→ rectifier effect

Screening effect (1919) S

(Maxwell-Lippmann effect)
If the conductivity of a body becomes infinitely large, Maxwell's second equation implies that the magnetic induction remains constant: any change in the external magnetic field produces an eddy current at the surface of the body, which persists because of the vanishing resistance. The eddy current balances out the induction change and the external field is screened off. This effect plays a role in superconductivity. It provides an explanation of the induction of persistent currents in rings or other twofold or manifold connected bodies.

MAXWELL, James Clerk
English physicist, 13 June 1831 – 5 November 1879, Professor in Aberdeen, London and Cambridge

LIPPMANN, Gabriel Jonas
French physicist, 16 August 1845 – 13 July 1921, Professor in Paris from 1883
Lippmann, G., *C.R. Paris* **168,** (1919)
Lit.: [30] (additional lit. see Table 44)

Seebeck effect

→ thermoelectric effects

Self-excitation

→ dynamoelectric principle

Self-focussing (1962) L

The light power level in the cavity of a laser can be so high that the refractive index of the active material is altered. Focussing of the radiation in the material may then occur, so causing damage. The effect is known as self-focussing and can also take place outside the cavity if the power flux density is high enough.
The effect is produced by the nonlinear part of the electrical polarization of the medium. This gives rise to a nonlinear dispersion relation, characterized by the fact that the waves become unstable under longitudinal but remain stable under transverse perturbations. The theory of this effect is extremely complicated. In the material, the wave front assumes a concave form which then leads to self-focussing.
→ nonlinear optical effects
Askaryan, *Zh. Eksp. Teor. Fiz.* **42**, 1567 (1962)
Lighthill, M. J., *J. Inst. Math. Its Appl.* **1**, 269 (1965)
Lit.: [32] (additional lit. see Table 31)

Senftleben effect K

The change in thermal conductivity of a gas in a magnetic field is called the Senftleben effect. The thermal conductivity in a magnetic field is dependent only on the ratio of the field strength to the pressure.
Lit.: [38] (additional lit. see Table 54)

Shape effect

→ magnetostriction

Shapiro's experiment (1964, 1968) R

In this experiment, transit times are measured by radar echo from Venus. The Earth-Venus-Earth transit time is measured and compared with the theoretical prediction. The general theory of relativity implies a transit time 10^{-4}s greater than

that given by Newton's theory. The experiments at present have an accuracy of 5 % and agree with Einstein's theory.
Shapiro, I. I., *Phys. Rev. Lett.* **13**, 789 (1964)
Shapiro, I. I., *Phys. Rev. Lett.* **20**, 1265 (1968)
Shapiro, I. I., *Phys. Rev. Lett.* **26**, 1132 (1971)
Lit.: [59] (additional lit. see Table 43)

Shapiro's experiment (1967) R

Shapiro used the radio emission from the quasars 3 C 273 and 3 C 279 in order to observe the *light deflection* produced by the gravitational field of the Sun in the radio range. As radio astronomy can now provide measurements of high accuracy, the deflection can be obtained with little error. The mean lies very close to the theoretical value (the ratio of measured to theoretical value is, for example, 1.04 ± 0.15, 0.99 ± 0.12, etc.).
→ light deflection
Lit.: [59] (additional lit. see Table 43)

Shearing effect

→ piezoelectric effect

Shot effect

(Johnsen effect, noise effect, fluctuation effect, small-shot effect)
This effect is due to the statistical fluctuations in an electron current under vacuum. It is more pronounced the smaller the number of electrons involved in the current. The fluctuation frequencies occuring in the shots theoretically cover the entire range from zero to infinity. The energy contained in a shot is uniformly distributed over the entire frequency band.
→ cooling effect, → Schottky effect,
→ Funkel effect
Schottky, W., *Ann. Phys.* **57**, 541 (1918)
Schottky, W., *Ann. Phys.* **68**, 157 (1922)
Schottky, W., *Phys. Rev.* **28**, 84 (1926)
Johnsen, J. B., *Phys. Rev.* **26**, 71 (1925)
Johnsen, J. B., *Phys. Rev.* **32**, 97 (1928)
Lit.: *dictionaries,* [50 Vol. 4, part 4]

Shpol'skii effect (1952) AQ II

With a certain class of compounds, very narrow fluorescent lines are obtained for molecules frozen at low temperatures. At such temperatures thermal motion is reduced. The Doppler and collisional line broadenings are eliminated, and natural line widths are attained. This is known as the Shpol'skii effect and has been used in the selective detection of aromatic hydrocarbons. The molecules are excited by monochromatic laser light and narrow-band fluorescence is observed.
Shpol'skii, E. B. et al., *Dokl. Akad. Nauk SSSR* **87**, 935 (1952)
Lit.: [42] (additional lit. see Table 21)

Shubnikov-de Haas effect SS

There are pronounced oscillations in the electrical resistance (static conductivity) of a single crystal at low temperatures in a strong magnetic field. The oscillations are of the same type as in the → de Haas-van Alphen effect.

SHUBNIKOV, Aleksei Vasilevich,
29 March 1887
Lit.: [1] (additional lit. see Table 51)

Size effects (1958, 1962) SS

All effects dependent on the size and shape of small bodies, for example, the dependence of resistivity on sample thickness are known as size effects. Quantum-mechanical effects occur when the de Broglie wavelength of the electrons is of the same order as the sample dimensions. Other effects occur when the electron mean free path is substantially greater than the sample size.
In the *Gantmacher* or *radio-frequency size effect,* the experimental conditions are similar to those in the → Azbel-Kaner effect. Measurements can be made on oscillations in the absorption and reflection or in the surface impedance of thin metal films in the MHz range as influenced by a homogeneous magnetic field and by temperature. This provides information on the Fermi surface of the material under investigation.
Gantmacher, V. F., *ZhETF* **43**, 345 (1962)
Kaner, E. A., *Dokl. Akad. Nauk SSSR* **119**, 471 (1958)
Lit.: [1, 75] (additional lit. see Table 51)

Skin effect EM

For high-frequency current the ohmic resistance of a wire is very much higher than that for direct current. The effective current cross section is reduced, since the current flows almost exclusively in a thin skin layer. The current lines in the inte-

rior are displaced to the surface because of induction.
The resistance is dependent on the square root of the quotient of frequency and conductivity. This applies for normal temperatures and not too high frequencies.
In practice, this means that the resistance of a conductor increases considerably with frequency, which plays an important role in high-frequency engineering.

ZENNECK, Jonathan,
German physicist, 15 April 1871 – 8 April 1959, Professor in Danzig, Brunswick, and Munich
Zenneck, J., *Ann. Phys.* (Leipzig) **11,** 1135 (1903)
Zenneck, J., *Textbook of Wireless Telegraphy* [in German], 4th Ed. Enke, Stuttgart (1916)
Lit.: [1, 6b, 50 Vol. 4, part 1] (additional lit. see Table 26)

Skin effect, anomalous (1940) EM

The skin layer involved in electrical conduction at high frequencies is very thin. At low temperatures, it can become less than the electron mean free path. In that case, the classical theory of electrical conductivity breaks down, and the exact solution is obtained from the transport equation: the effective conductivity is independent of the mean free path and is proportional to the reciprocal of the cube root of the frequency. This is called the anomalous skin effect.
The anomalous skin effect can be observed only above a certain limiting frequency and at low temperatures.
→ Azbel-Kaner effect, → de Haas-van Alphen effect
London, H., *Proc. R. Soc. Ser A* **176,** 522 (1940)
Lit.: [1, 6d, 71] (additional lit. see Table 26)

Slope effect

→ Barkhausen effect in thin films

Small-shot effect

→ shot effect

Smith-Purcell-Salisbury effect (1953) AQ II

If a flat electron beam accelerated to a few hundred keV strikes a metal grating, almost coherent radiation is produced at a defined angle of less than 90° (Smith and Purcell 1953).
In 1953, Salisbury independently reported a similar experiment. He explained the emission by means of the electrostatic theory of charge patterns: in a metal, positive charges follow the electrons as patterns. The amplitude of the metal lattice produces oscillations in the electrons and charge patterns, so leading to the emission of dipole radiation.
Smith, S. J. and Purcell, E. M. *Phys. Rev.* **92,** 1069 (1953)
Salisbury, W. W., *U. S. Patent No. 2634372*
Lit.: [87 Vol. XXIX] (additional lit. see Table 21)

Smoluchowski effect (1911) TK

The thermal conductivity of a gas is normally independent of the pressure, but if the molecular mean free path is of the same order as the dimensions of the vessel, the conductivity increases with pressure. If the vessel is filled with an extremely fine powder, the thermal conductivity is reduced even at normal pressure. This is called the Smoluchowski effect and is used in the production of containers with good thermal insulation, such as Dewar vessels.

SMOLUCHOWSKI, Maryan (Marian), Ritter von Smolan,
Austrian-Polish physicist, 28 May 1872 – 5 September 1917
Smoluchowski, M., *Ann. Phys.* **35,** 22 (1911)
Lit.: [72] (additional lit. see Table 54)

Snoek effect (1938, 1941) So

Small amounts of carbon have pronounced effects on the plastic behavior of iron. The Snoek effect is the preferential occupation by carbon of sites on one of the three faces in the cubic lattice of iron. In general, the name Snoek effect is now given to the internal friction arising from the occupation of interstitial lattice sites in a cubic body-centered crystal by foreign atoms such as C, O, and N. This occupation produces a local reduction in the symmetry. Under mechanical stress, the degeneracy is lifted again, which results in an energy loss, i.e., the excitation of thermal oscillations.
Snoek, J. L., *Physica* **5,** 663 (1938)
Snoek, J. L., *Physica* **6,** 591 (1939)
Snoek, J. L., *Physica* **8,** 711 (1941)
Lit.: [37] (additional lit. see Table 50)

Sondheimer effect (1950) SS

If there is an angle between a microwave field and a magnetic field i.e. different to the → Azbel-Kaner effect), nonresonant size effects result. In the Sondheimer effect, the electrons move along circular paths, whose axes form an angle with the surface of the sample. This produces a change in the static conductivity. The effect also offers scope for measuring the curvature of the Fermi surface at its boundary points. In conjunction with the Azbel-Kaner effect, the Fermi velocity can be measured at the boundary points. By changing the magnetic field orientation, this can be done for almost the entire Fermi surface.

SONDHEIMER, Ernst Helmut,
German physicist, 8 August 1923
Sondheimer, E., *Phys. Rev.* **80,** 401 (1950)
Lit.: [1] (additional lit. see Table 51)

Spectral selective photoelectric effect

→ photoelectric effect

Staebler-Wronski effect (1977) SCC

The Staebler-Wronski effect describes the change (usually a reduction) in the dark conductivity and photoconductivity of hydrogenated amorphous silicon (a-Si:H) following or during illumination ($h\nu > E_G$). The magnitude of the effect is a function of the light intensity times time. The effect is reversible, i.e., the dark conductivity and photoconductivity return to their original values after annealing (about 150 ° C for about 30 min). It is now assumed that the light produces electron-hole pairs, which recombine in the a-Si:H; the released recombination energy produces metastable states in the form of dangling bonds, which are free valencies that may arise by the breaking of weak Si-Si links in the amorphous framework. These new states act as recombination centers for subsequent carriers.
(However, there are also other model representations for the mechanism of the Staebler-Wronski effect.)
In a solar battery, the main influence from the Staebler-Wronski effect is a reduction in the short-circuit current (photocurrent). Also, a-Si:H solar batteries are reversibly restored by annealing. At present, it appears impossible to produce completely stable a-Si:H cells.

Staebler, D. L. and Wronski, C. R., *Appl. Phys. Lett.* **31,** No. 4, 292 (1977)
Lit.: [100, 110, 114]

Stark effect (1913) Eo

A strong electric field splits spectral lines in the same way as in the → Zeeman effect to give several components. For hydrogen, there is a linear dependence on the field strength (*linear Stark effect*), whereas all other atoms and molecules exhibit a quadratic dependence (the *quadratic Stark effect*). The degeneracy in the energy levels having the same principal quantum number is lifted, and the spectrum contains the number of lines corresponding to the orbital momentum quantum number. The effect does not have the same significance as the Zeeman effect. However, the Stark effect must be borne in mind if one wishes to calculate the energy terms for foreign atoms in host lattices. The atomic fields produce term splitting.
→ Jahn-Teller effect

STARK, Johannes,
German physicist, 15 April 1874 – 21 June 1957, Professor in Hannover, Aachen, Greifswald and Würzburg, 1933 – 1939 President of the State Technical Physics Institute
Stark, J., *Sitz.-Ber. Preuss. Akad. der Wiss.* Vol. 2, 932 (1913)
Stark, J., *Ann. Phys.* **43,** 83 (1914)
Stark, J., *Ann. Phys.* **48,** 26 (1915)
la Surdo, A., *Atti Acad. R. Lincei* **22,** 664 (1913)
Lit.: [6c, 6d, 10, 27, 50 Vol. II, 2.2] (additional lit. see Table 27)

Stark-Doppler effect

→ optical Doppler effect

Stavermann effect TK

The osmotic pressure of a solution is lower than that calculated from the total molar concentrations when the solution contains components that affect membrane permeability.
Lit.: [75] (additional lit. see Table 54)

St. Elmo's fire

→ Elmo's fire

Stern-Gerlach experiment (1921) AQ II

(Stern-Gerlach effect)

An atom with a magnetic moment placed in an inhomogeneous magnetic field experiences a force in the direction of the inhomogeneity. In this way, an atomic beam, for instance, can be deflected from a straight line, which allows the magnetic moments of the individual atoms to be measured. In addition, so-called direction quantization is observable, in which a beam is split into a number of partial beams equal to the number of positions the spin can take with respect to the magnetic field. If J is the quantum number of the resulting angular momentum of the atom, then $(2J + 1)$ partial beams are possible. The number of partial beams enables one to determine the total angular momentum of the corresponding atomic state. The *atomic beam resonance method* for measuring nuclear magnetic moments is based on the Stern-Gerlach experiment.

STERN, Otto,
German physicist, 17. February 1888 – 17 August 1969,
Professor in Frankfurt a.M., Rostock, Hamburg, from 1933 in Pittsburg (USA)

GERLACH, Walther,
German physicist, 1 August 1889 Biebrich – 15 August 1979 Munich,
Professor in Frankfurt, Tübingen and Munich

Stern, O., *Z. Phys.* **7**, 249 (1921)
Gerlach, W. and Stern, O., *Z. Phys.* **8**, 110 (1921)
Gerlach, W. and Stern, O., *Z. Phys.* **9**, 349, 353 (1922)
Stern, O., *Ann. Phys.* **74**, 673 (1924)

Lit.: [6d, 7, 16, 17, 22d, 41, 60, 66, 69, 70 Vol. II] (additional lit. see Table 21)

Stimulated emission

→ lasers

Stress birefringence

→ birefringence

Stretching birefringence

→ birefringence

Stroboscope effect G

The light flux from an electric lamp driven by alternating current shows periodic variations with twice the mains frequency. This light flickering at 100 Hz from 50 Hz supply is normally not observed, since the eye cannot detect variations with this frequency. However, under certain conditions perturbing phenomena can occur when rapidly moving objects are illuminated with such light. For example, a rapidly turning part of a machine may appear to be at rest if its rotational frequency is equal to the flicker frequency. This is known as the stroboscope effect. This effect is used to observe fast movements. The frequency of the lamp (generally a flash lamp) is adjusted until the motion comes to rest or becomes very slow. For example, the backward rotation of wheels in a film is derived from the stroboscope effect.

Lit.: [16] (additional lit. see Table 19)

Substituent effects AQI

(perturbation propagation effect, F effect, field effect, field (F) effect, I effect, induction effect, mesomeric effect)

The propagation of a perturbation in the electron distribution of a molecule is called a substituent effect. If the perturbation involves dipole induction along a molecular chain, one speaks of the *induction or I effect*. If the perturbation is produced directly by the field, the *field (F) effect* results.

In unsaturated compounds, there are additional *mesomeric substituent effects,* in which the electrons of a substituent influence the electron distribution in the rest of the molecule via interaction with the electrons in the unsaturated bond.

Lit.: [75] (additional lit. see Table 20)

Suhl effect (1949) SC

The motion of injected carriers in a semiconductor is influenced by the joint action of external electric and magnetic fields. For example, carriers can be brought into regions of elevated recombination close to the crystal surface. The current measured at a collector is dependent on both fields. The effect is used to examine the recombination characteristics of semiconductors. Also, it is of use in a magnetic diode, which can act as an amplifier or multiplier.

SUHL, Harry
German-American physicist, 18 October 1922

Suhl, H. and Shockley, W., *Phys. Rev.* **75**, 1617 (1949)

Lit.: [75, 87 Vol XX] (additional lit. see Table 48)

Superconductivity (1911) S

In 1911, during research into the temperature dependence of electrical conductivity at low temperatures, Heike Kamerlingh Onnes discovered the phenomenon of superconductivity: within a temperature range of a few hundredths of a degree around 4.2 K, the resistivity of mercury became zero. The relative resistance change ran over several powers of 10, from about 10^{-1} to less than 10^{-5}. The result was surprising and demonstrated a new state of matter. Today, resistance changes of up to 10^{-12} are found. The effect involves hysteresis (→ hysteresis effect). Superconductivity is a low-temperature phenomenon, as is superfluidity (→ Onnes effects). Various other effects are related to it, and provide an insight into the conduction mechanism in superconductivity.

At low temperatures, the electrons are coupled in *Cooper pairs* with the following properties: the two electrons in a Cooper pair have equal but opposite momentum and opposite spins (inherent) angular momentum). A Cooper pair therefore has a total momentum and spin of zero. The properties of superconductivity are controlled by the existence of Cooper pairs (Cooper effect).

The superconducting state can be eliminated not only by a rise in temperature but also by an external magnetic field. The superconduction domain can be demonstrated in a field-temperature diagram, with the critical magnetic field quadratically dependent on the temperature. The diagram is usually called a phase diagram. One distinguishes between normal-conducting and superconducting phases.

Not only metals but alloys and intermetallides exhibit superconductivity, so the following classification results:

Superconductors of the first kind: (soft superconductors)

Pure crystalline elements, sudden and reversible transition to the normal conducting state at low magnetic field strengths. An external magnetic field penetrates only into a thin surface layer. (Te, Pb, Ta, Hg, Sn, In, Al, Ta and Tl)

Superconductors of the second kind:

The elements Nb and V, alloys such as lead-bismuth and niobium-zirconium, and intermetallides such as Nb_3Sn or V_3Ga. These materials show a division into normal and superconducting domains. This leads to an increase in the critical flux density and current density. When a current is flowing the domains can be moved by an external magnetic field.

Superconductors of the third kind: (hard superconductors)

In these superconductors, the domain movement is hindered by lattice defects.

Superconductors of the first kind exhibit no galvomagnetic, thermoelectric, or thermomagnetic effects. The Hall effect can be observed in superconductors of the second and third kinds.

This is related to the occurrence of superconducting and normal-conducting domains.

The other effects associated with superconductivity are listed in Table 44 (S) at the end of this book.

→ Onnes effect

KAMERLINGH ONNES, Heike,
Dutch physicist, 21 September 1853 – 21 January 1926, from 1882 Professor in Leiden

Kamerlingh Onnes, H., *Commun. Kamerlingh Onnes Lab. Univ. Leiden* **120b,** (1911)
Kamerlingh Onnes, H. *Commun. Kamerlingh Onnes Lab. Univ. Leiden Suppl.* **34,** (1913)
Wonn, H., *Phys. Status Solidi* **8,** 3, 639 (1965)
Lit.: [6d, 11, 15, 22d, 44], *dictionaries*

Superelasticity

→ memory effect

Switching effects

→ Ovshinsky effect

Synergetic phenomena Sy

Synergetics is a new interdisciplinary field in physics, biology, chemistry, economics, population behavior, etc. the purpose being to describe complex systems involving the cooperative behavior of numerous subsystems and exhibiting ordered structures far from thermodynamic equilibrium.

Nonlinearity is common to all these phenomena. The mathematical description is based on Thoms's catastrophe theory or master equations such as the Fokker-Planck equation.

Synergetics provides an understanding of self-organization in biological systems and evolution. Sociological processes, consumption, or voting behavior can be described quantitatively.

The classical physical examples are → lasers, magnetism, flow phenomena (→ Benard effect, → Taylor effect), and plasma instabilities.

Haken, H.: *Synergetics*. Springer, Berlin-Heidelberg-New York 1983
Lit.: [102a, 103b, 112a, 116a]

Szilard-Chalmers effect (1934) N

The name Szilard-Chalmers effect is given to a change in the state of chemical bonding in radioactive atoms upon recoil following neutron capture and subsequent particle emission (e.g., gamma rays).
The atoms generally undergo considerable recoil, so that they can escape from the molecular lattice. The effect is used in the production of radioactive isotopes.

SZILARD, Leo,
Hungarian-American physicist, 11 February 1898 – 30 May 1964, from 1938 in the USA, after 1946 Professor in Chicago
Szilard, L. and Chalmers, T. A., *Nature* **134,** 462 (1934)
Lit.: [75] (additional lit. see Table 38)

Taylor effect (1923) Sy

In Couette flow between rotating cylinders, instabilities arise at a critical angular velocity (Taylor instability), in which periodic eddies along the axis form a regular pattern. As the Taylor number (dimensionless) of the flow increases, first a wave-type flow results and then turbulence.
→ synergetic phenomena
Haken, H.: *Synergetics*. Springer, Berlin-Heidelberg-New York 1983
Lit.: [102a, 103b, 112a, 116a]

Telescope effect So

When a high-polymer fiber is stretched, an increase in hardness is observed. This leads to necking, which arises at the beginning of yield without leading to failure when the fiber is loaded. The telescope-type expansion of the neck is why the process is known as the telescope effect. The cause of the yield is the kinking of the chains in the unstretched form, which on stretching disappears. In such an oriented body, the external forces are taken up by the principal valencies, which are by an order of magnitude stronger than the secondary valencies. This leads to an increase in strength at small cross sections.
Lit.: [90] (additional lit. see Table 50)

Texture-conversion effect

→ liquid-crystal effects

Thermal aftereffect

→ magnetic aftereffects

Thermal diffusion

→ diffusion

Thermal diffusion potential

→ thermoelectric effects,
→ thermoelectric homogeneous effect

Thermal effusion

→ Knudsen effect

Thermistor effect (1939) SCC

In a semiconductor, the electrical conductivity is very much dependent on temperature. The temperature coefficient of the resistivity of a semiconductor can be positive or negative, and is much larger than that in metals.
A thermistor (hot conductor) is a semiconductor that has a very pronounced negative temperature coefficient (NTC thermistor). In this case, a rise in temperature causes additional electrons to be released, which produce a strong exponential reduction in the resistance as the temperature increases. Such a semiconductor consists of sintered mixed ceramic crystals, mainly oxides of transition metals belonging to group III of the periodic table. The current-voltage characteristic is of pronounced nonlinear type. Hot conductors are used in the measurement and control of devices and systems.
Sachse, H., *Wiss. Veröff. Siemens*, Paper 19, 214 (1939)
Lit.: [84 Vol. I, 91, 92] (additional lit. see Table 49)

Thermodynamic pressure effect

→ Onnes effect

Thermoelectric effects Te

This term covers the following effects:

Seebeck effect	Thermoelectric inhomogeneous effects
Peltier effect	
Thomson effects	Thermoelectric homogeneous effects
Benedicks effects	
Bridgman effect	(internal Peltier effect)
Phonon-drag effect	(lattice Peltier effect)

→ current conduction effects, → galvanomagnetic effects, → thermomagnetic effects

Benedicks effects (1916/1918)
A thermo-EMF arises in a homogeneous conductor when there is an extremely large temperature gradient along it *(first Benedicks effect)*. The *second Benedicks effect* is an extension of the → Thomson effect: a homogeneous conductor carrying a current shows a temperature rise (very small) proportional to the current at a "restriction". The effect has a certain similarity to the → Knudsen effect (thermal effusion).
In metallic conductors the electrothermal homogeneous effects are very small in contrast to other effects, but allowance must be made for them in metal-semiconductor combinations. The order of magnitude is a few microvolts per degree.
BENEDICKS, Carl Axel Frederic,
27 May 1875 – 16 July 1958 Stockholm
Benedicks, C. A. F., *Ann. Phys.* **55**, 1, 103 (1918)
Benedicks, C. A. F., *Ann. Phys.* **62**, 185 (1920)
Benedicks, C. A. F., *C. R.* **167**, 296 (1918)
Benedicks, C. A. F., *Math. Astron. de Phys.* **23A**, No. 27
Benedicks, C. A. F., *Math. Astron. de Phys.* **24A**, No. 1, 7 (1933)
Lit.: [6b, 22b, 22d, 30, 75, 76, 78, 88] (additional lit. see Table 53)

Bridgman effect (1925)
(internal Peltier effect)
In the internal Peltier effect heat *(Bridgman heat)* is produced in an anisotropic metal when a current flows, and in particular when the current changes its direction in relation to the crystal axes. The experimental results show that a thermo-EMF also occurs between differently oriented crystals of the same metal.
BRIDGMAN, Percy Williams,
American physicist, 21 April 1882 – 20 August 1961, Professor at Harvard University
Bridgman, P. W., *Proc. Nat. Acad. Sci. USA* **11**, 608 (1925)
Bridgman, P. W., *Proc. Am. Acad. Arts Sci.* **61**, 101 (1925)
Bridgman, P. W., *Phys. Rev.* **39**, 702 (1932)
Bridgman, P. W., *Phys. Rev.* **42**, 858 (1932)
Lit.: [30, 74, 75] (additional lit. see Table 53)

Peltier effect (1834)
(electrothermal inhomogeneous effect)
When a current flows in a circuit made of two different homogeneous conductors, the junctions receive or lose heat in accordance with the current direction *(Peltier heat)*.
The Peltier effect is the converse of the → Seebeck effect. The amount of heat produced or absorbed is directly proportional to the current strength, and the coefficient of proportionality is known as the Peltier coefficient. The Seebeck and Peltier effects both occur in modern semiconductors, where they are particularly pronounced, and they are used in components in the conversion of thermal energy to an electrical form and in cooling components to attain low temperatures.
PELTIER, Jean Charles Athanasa,
French physicist, 22 February 1785 – 27 October 1845, Paris
Peltier, J. Ch. A., *Ann. Chim. Phys.* **56**, 371 (1834)
Lit.: [3, 6b, 22d, 23, 27, 30, 38 Vol. VIII, 43, 49, 53, 69, 70] (additional lit. see Table 53)

Phonon-drag effect
(Gurevich effect)
In a metal electrons are scattered at phonons and vice versa. When a current flows in a metal, this corresponds to a shift in the Fermi surface. Due to the electron-phonon interaction, the phonon Fermi surface is also displaced. The two systems bry to attain an equilibrium.
The phonon flux corresponds to the heat flux released by the electric current (→ Peltier effect). However, at normal temperatures, the phonon-phonon interactions are generally so large that it is difficult to observe this heat flux *(Gurevich effect)*.
The phonon-drag effect is readily observed at low temperatures. The direction in which the phonons are scattered at the electrons relative to the external field determines whether the current is reduced by the additional phonons or whether the phonons are emitted in the opposite direction to the electrons.
The lattice Peltier effecte can therefore be positive or negative.
→ drag effect, → photon-drag effect
GUREVICH, Anolde,
Russian-American physicist, 12 June 1911
Lit.: [2, 71] (additional lit. see Table 53)

Seebeck effect (1822)
(thermoelectric inhomogeneous effect)
A circuit composed of two different homogeneous conductors produces a thermo-EMF when the junctions are at different temperatures. The thermo-EMF is proportional to the temperature difference, and the coefficient of proportionality is called the differential *thermoelectric force*.
The thermo-EMF (or thermocurrent) is produced because the charge carriers take up a new distribution on account of the temperature difference.
SEEBECK, Thomas Johann,
German physicist, 9 April 1770 Reval – 10 December 1831 Berlin
Seebeck, T. J., *Abh. Kgl. Akad. Wiss. Berlin* **23**, 265 (1822)
Seebeck, T. J., *Ann. Phys.* (1) **73**, 430 (1823)
Seebeck, T. J., *Ann. Phys.* (2) **6**, 1, 133 (1826)
Lit.: [3, 6b, 22b, 22d, 30, 86] (additional lit. see Table 53)

Table 15. Thermoelectric effects (see also Tables 23, 30, 53 and 55)

	yields	Name of effect		
Thermoelectric inhomogeneous effects				
Temperature difference between two contact points in different conductors	Thermo-EMF	Seebeck effect	Confirmed by experiment	
Current flowing through different conductors	Temperature difference between contacts	Peltier effect	Confirmed by experiment	
Thermoelectric homogeneous effects				
Temperature gradient along conductor	Thermo-EMF	First Benedicks effect	Almost confirmed by experiment	*First order effects*
Constant temperature along conductor with a constriction (reduced cross section) and current flowing through it	Temperature gradient along conductor	Second Benedicks effects	Not definitely observed	
Temperature gradient along conductor and currunt flowing through it	Change in temperature gradient	Thomson effect	Confirmed	*Second order effects*
Varying temperature gradient	Thermo-EMF	Inverse Thomson effect	Not yet observed	
Thermoelectric effects in crystals				
Current direction different from crystal axes	Heat	Bridgman effect	Observed	
Current flowing in a metal	Phonon flux	Peltier effect in lattice, Gurevich effect, phonon-drag effect	Observed	

Thermoelectric homogeneous effect
In general, a temperature difference in a mixture of substances in the steady state produces partial unmixing of the components *(thermal diffusion)*. With electrolytes, however, the slightest ionic unmixing produces an electric field, which opposes further separation. In this case, thermal diffusion leads to potential differences *(thermal diffusion potential)* in virtually homogeneous substances.
Lit.: [76] (additional lit. see Table 53)

Thomson effect (1856)
A current flowing in a homogeneous conductor with a temperature gradient is accompanied by the production or absorption of heat in addition to Joule heat. If the current flows in the direction of increasing temperature, heat is produced if the coefficient is positive and vice versa. If the coefficient is negative, the relationships are reversed. The heat produced is proportional to the current and the temperature difference. The Thomson effect is increased by the → Joule effect (Joule heat), which is quadratically dependent on the current.
→ galvanomagetic effects

THOMSON, William, Lord Kelvin

Thomson, W., *Math. Phys. Papers* **2**, 192 (1882)
Thomson, W. *Philos. Trans. R. Soc.*, London **146**, 663 (1856)
Thomson, W., *Math. Phys. Papers* **2**, 207 (1884)
Thomson, W., *Math. Phys. Papers* **1**, 232, 266 (1882)
Lit.: [100, 114, 117]

Thomson effect, inverse (1856)
For reasons of symmetry, there must be an inverse Thomson effect, in which a changing temperature gradient produces a thermo-EMF. This effect has not yet been observed. Both Thomson effects are second-order thermoelectic effects.
Lit.: [30] (additional lit. see Table 53)

Thermoelectric homogeneous effect
→ thermoelectric effects

Thermoelectric inhomogeneous effect
→ thermoelectric effects

Thermoionic effect
→ Langmuir effect

Thermoluminescence effect
→ luminescence effects

Thermomagnetic effects Tm, CC
This term covers effects occurring in conductors when a magnetic field is applied in the presence of a heat flux: the Righi-Leduc effect, the Ettingshausen-Nernst effect, the change in thermal resistance in a magnetic field, the thermo-EMF (Seebeck effect) between magnetized and unmagnetized parts of the same material (second Ettingshausen-Nernst effect).
→ current conduction effects

Ettingshausen-Nernst effect (1886)
This ist the converse of the Ettingshausen effect: a transverse potential difference is produced in a transverse magnetic field when there is a transverse heat flux. The effect is the thermal analogue of the → Hall effect.
The electric current is replaced by the heat flux and the Hall voltage is replaced by the temperature difference.
The effect can be used in a detector of thermal radiation.
→ OEN effect
The *second Ettingshausen-Nernst effect* is the formation of a thermo-EMF between magnetized and unmagnetized parts of the same conductor.

Nernst, G., *Wied. Ann.* **24**, 343 (1886)
Nernst, G., *Wied. Ann.* **31**, 760 (1887)
Elbel, T. and Soa, E. A., *Feingerätetechnik* **28**, 243 (1979)
Nernst, G., *Wiener Anz.* Nr. VIII (1886)
Lit.: [94, 100, 104, 110, 114]

Maggi-Righi-effect
(second Righi-Leduc effect)
A homogeneous conductor carrying direct current in a transverse magnetic field experiences a longitudinal temperature difference or a change in thermal conductivity.

MAGGI, Gian Antonio,
Italian physicist, 19 February 1856 – 1937

RIGHI, Augusto,
Italian physicist, 27 August 1850 – 8 Juni 1920

OEN effect
(optically induced Ettingshausen-Nernst effect)
In addition to the photo-Hall effect, there is an optically induced Ettingshausen-Nernst effect:

radiation absorption in a suitable semiconductor wafer results in a temperature difference between its front and back sides. This leads to a heat flux. In a transverse magnetic field, a voltage is also produced, which is perpendicular to the beam direction and to the magnetic field. The effect is used as a radiation detector.

Motschmann, H., *Siemens Forschungs- und Entwicklungsberichte* **3**, 19 (1974)

Lit.: [94, 100, 104, 110, 114]

Table 16. Galvanomagnetic and thermomagnetic effects (see also Tables 23, 30, 53 and 55)

	Galvanomagnetic effects		Thermomagnetic effects	
	Electric current *J* flows through the conductor		Heat flux *W* passes through the conductor	
	A potential difference arises	Temperature difference	A temperature difference arises	Potential difference
Transverse *B* longitudinal effects	Hall effect Transverse potential difference 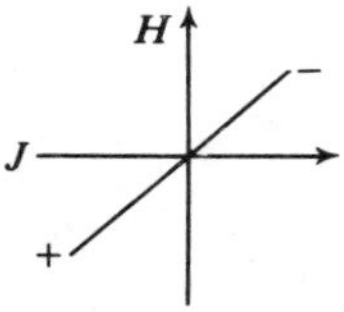 H, J, +, – (1)	Ettinghausen effect Transverse temperature difference 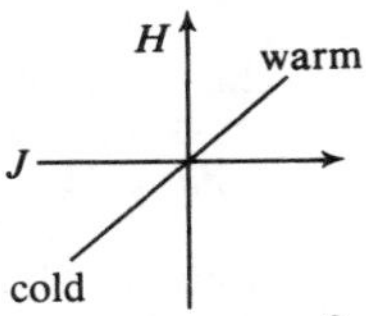 H, J, warm, cold (2)	First Righi-Leduc effect Transverse temperature difference 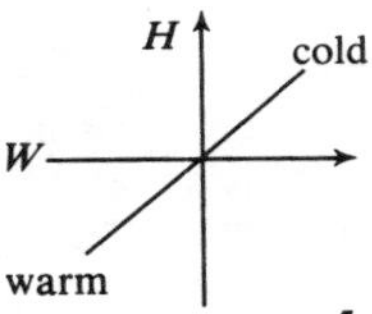 H, W, cold, warm (5)	First Ettingshausen-Nernst effect Transverse potential difference 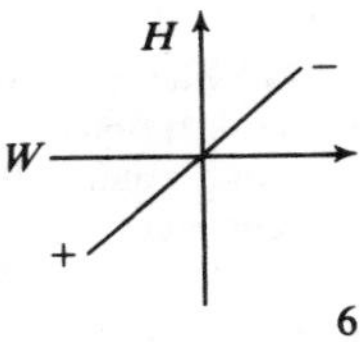 H, W, –, + (6)
Transverse *B* transverse effects	Thomson effect Change in longitudinal potential difference = Change in resistivity = Transverse magnetoresistance 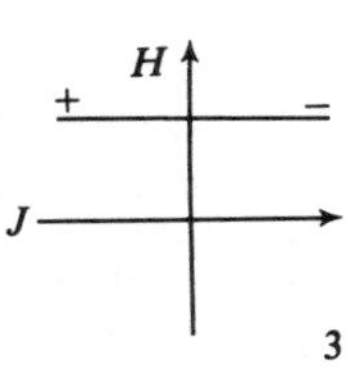 H, +, –, J (3)	Nernst effect Longitudinal temperature difference = Peltier effect between magnetized and unmagnetized materials 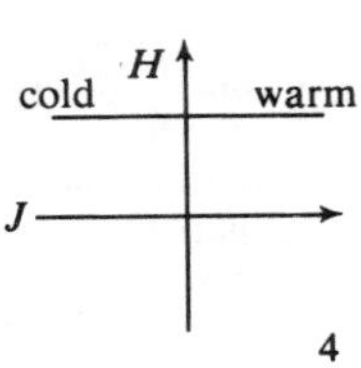 cold, H, warm, J (4)	First Righi-Leduc effect Maggi-Righi-Leduc effect Change in thermal resistance (conductivity) = Transverse thermal magnetoresistance 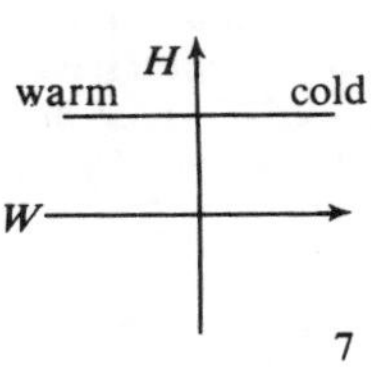 warm, H, cold, W (7)	Second Ettingshausen-Nernst effect Longitudinal potential difference, thermo-EMF between magnetized and unmagnetized materials 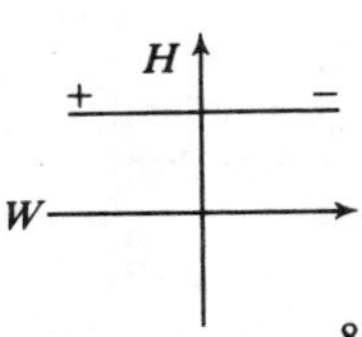 H, +, –, W (8)

Table 16 (Fortsetzung)

	9	10	11	12
Longitudinal B longitudinal effects	Change in resistivity = Longitudinal magneto-resistance; 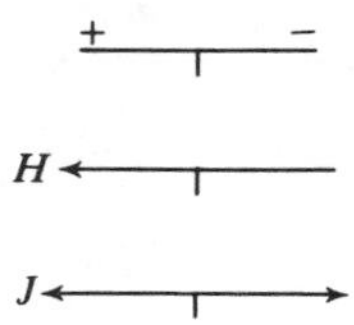 + −; H ←; J ↔	Peltier effect between magnetized and unmagnetized material; 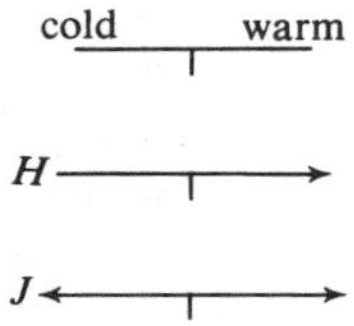 cold warm; H →; J ↔	Change in thermal resistance (thermal conductivity) = Longitudinal thermal magneto-resistance; 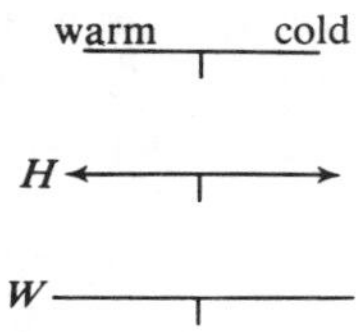 warm cold; H ↔; W —	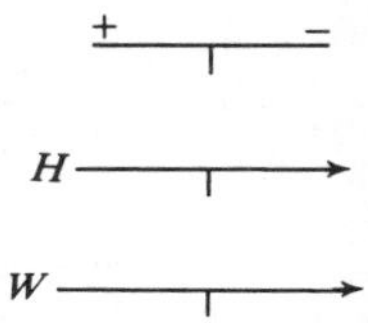 (Seebeck effect) Thermo-EMF between magnetized and unmagnetized materials; + −; H →; W →

Righi-Leduc effect (1887)
If the directions of a heat flux and a magnetic field in a homogeneous conductor are mutually perpendicular, a temperature gradient perpendicular to them both is set up.

LEDUC, Sylvestre Anatole,
French physicist, 22 April 1856 – 15 April 1937
Leduc, S. A., *C. R.* **104**, 1783 (1887)
Leduc, S. A., *C. R.* **105**, 250 (1887)
Righi, A., *Mem. R. Acc. dei Lincei, Rendic.* (4) **3**, 6 (1887)
Lit.: [94, 100, 104, 110, 114]

Thermooptic effect
→ liquid-crystal effects

Thirring-Lense effect (1918) R
According to the general theory of relativity, a gyroscope turning in a rotating hollow body should show precession. As the effect is very small, it has so far not been observed. It appears possible that it might be detected by means of satellite experiments.

THIRRING, Hans,
Austrian physicist, 23 March 1888 – 22 March 1976
LENSE, Joseph,
28 October 1890 Vienna
Thirring, H., *Phys. Z.* **19**, 3, 10, 33, 204 (1918)
Thirring, H., *Phys. Z.* **22**, 1, 29 (1921)
Thirring, H. and Lense, J., *Phys. Z.* **19**, 8, 156 (1918)
Lit.: [40 Vol. II, 59] (additional lit. see Table 43)

Thomson effect
→ galvanomagnetic effects,
→ thermoelectric effects

Thomson effect, inverse
→ thermoelectric effects

Thomson scattering
→ scattering effects

Throttle effect
→ Joule-Thomson effect

Time effects R
(Hafele-Keating experiment 1971)
In this experiment the time dilation predicted by special and general relativity theory was measured. This became possible with the development of atomic clocks, which possess an accuracy of greater than 10^{-11}. Hafele and Keating flew in an

aircraft in easterly and westerly directions and compared four atomic clocks on board with the stationary standard clock before and after the flight. The results are given in Table 17.

Table 17. Results of the Hafele-Keating experiment

	Theoretical effects caused by: Gravitation ns	Velocity ns
Westerly flight	179 ± 18	96 ± 10
Easterly flight	144 ± 14	− 184 ±
	Theoretical overall effect ns	Experimental overall effect ns
Westerly flight	275 ± 21	273 ± 7
Easterly flight	− 40 ± 23	− 59 ± 10

The relativistic time effects were thus confirmed by the experiment.

Lit.: [59] (additional lit. see Table 43)

Touschek effect

In electron storage rings, Møller scattering leads to the loss of two electrons. At low energies (with a dependence of approx. E^{-5}), this restricts the maximum particle concentration in the circulating electron bunches and therefore the maximal event rate (in general very low).

→ particle scattering (Mott scattering)

(See also table 21 and 47)

TOUSCHEK

Austrian physicist 1921 – 1978

Touschek, *Phys. Rev. Letters* **10**, 407 (1963)

Track adaptation effect Me

(coincidence effect)

If a sound wave is incident on a wall at an angle, bending oscillations will propagate in the wall. If these correspond to the free bending oscillations of the wall, there is a coincident limiting frequency, at which the sound insulation is minimal.

Lit.: [85] (additional lit. see Table 35)

Transistor effect (1948) SCC

Bardeen, Brattain, and Shockley discovered the transistor effect during research on defect-electron motion in semiconductor crystals. They used a single germanium crystal bearing metal electrodes and injected charge carriers, while a second metal electrode was used to measure the current through the wafer as a function of the separation between the two electrodes. A basic current thus flowed through the wafer. They found that with correct positioning of the electrodes and with appropriate voltages, the current was amplified by a large factor: the carriers from the emitter diffused through a virtually field-free zone, the base, and were taken up by the collector, thus increasing the collector current. The transfer from point or whisker transistors to the now usual PNP or NPN transistors was very rapid.

Transistors have now virtually replaced vacuum tubes, apart from special uses, and they are the starting point for microelectronics, microprocessors, data processing, and so on.

BARDEEN, John,

American physicist, 23 May 1908,

from 1938 Professor at the University of Minnesota, 1945 – 1951 at Bell Telephone Laboratories, from 1951 Professor at the University of Illinois

BRATTAIN, Walter Houser,

American physicist, 10 February 1902 Amoy (China),

from 1929 at Bell Telephone Laboratories, 1967 Professor in Walla Walla (Wash.)

SHOCKLEY, William Bradford,

British-American physicist, 13 February 1910 London, worked at the Bell Telephone Laboratories, from 1963 Professor at Stanford University

Bardeen, J. and Brattain, W. H., *Phys. Rev.* **74**, 230, 231 (1948)

Bardeen, J. and Brattain, W. H., *Phys. Rev.* **75**, 1208 (1949)

Shockley, W. B., *Bell Syst. Tech. J.* **28**, 435, 2931 (1949)

Lit.: [6b, 22b, 22d, 28, 53, 74, 75, 91, 92] (additional lit. see Table 49)

Transition effect

→ Paschen-Back effect

Transit-time effects G

These are effects that occur as a result of particle or wave transit times. The electron transit time is relevant, for example, to high-frequency valves

or in the amplification of high frequencies by means of valves.
It is a property of electromagnetic waves that pulses, for example light pulses, alter their width and shape with the transit time in a dielectric material.
→ dispersion effects
Lit.: *Textbooks, dictionaries,* literature on high-frequency engineering

Transport effects G

(kinetic effects)
This term covers numerous irreversible phenomena that can be described by the kinetic theory of matter.
It includes viscosity, thermal conduction, → diffusion, the thermomechanical effect, the pressure effect, → diffusion thermal effects, → Knudsen effect, → thermal effusion, → Ludwig-Soret effect, electrical conductivity, → galvanomagnetic effects, → thermomagnetic effects, → thermoelectric effects, → radiometer effect.
The phenomena can be described by a modification of Boltzmann's transport equation adapted to the various circumstances. The name → cross effects is given to coupling between several transport effects.
Lit.: [20]

Transverse barrier-layer photoelectric effect

→ barrier-layer photoelectric effect

Transverse Doppler effect

→ optical Doppler effect

Transverse effect

→ piezoelectric effect, → Zeeman effect

Triboluminescence

→ luminescence effects

Trouton-Noble experiment (1903) R

If one assumes that the Galilean relativity principle and electrodynamics are valid, a moving dipole should exert a torque on itself. The motion of a charged parallel-plate capacitor shows no torque, which agrees with the results from Einstein's special relativity theory. Therefore, the Galilean relativity principle cannot be correct in electrodynamics. The Lorentz transformation gives the correct results. In this, the spatial coordinates and time are treated as equivalent, and the constancy of the speed of light is assumed.

TROUTON, Frederik Thomas,
British physicist, 1863 – 21 September 1922
Trouton, F. T. and Noble, H. R., *Proc. R. Soc.* **72**, 132 (1903)
Trouton, F. T. and Noble, H. R., *R. Soc. London, Philos. Trans.,* **202**, 18 (1904)
Lit.: [40] (additional lit. see Table 43)

Tunnelling effect (1928) AQ I

(quantum-mechanical tunnelling effect)
According to classical physics, a particle cannot pass through a potential hill of finite height and thickness. According to the laws of quantum mechanics, there is a precisely definable probability of tunnelling through the potential barrier. This occurs because the wave function of the particle must be continued uniformly under the potential barrier and in the outer region.
Therefore, the particle has a certain probability of being present in the potential barrier, which decreases as the thickness of the barrier increases. On account of wave-particle dualism, this applies also to waves.
The tunnelling effect gives an elementary explanation for nuclear α decay (Gamow, Condon, and Gurney) and is of considerable significance for semiconductor components such as tunnel diodes and Zener diodes.
In acoustics and optics, wave tunnelling is also observed, as well as in superconductivity.
→ Goss-Hähnchen effect, → shock effect, → Josephson effects, → electron-tunnelling effect, → Zener effect, → avalanche effect

GAMOW, George Anthony,
Russian-American physicist, 4 March 1904 – 19 August 1968, from 1934 Professor in the USA
Gamow, G., *Z. Phys.* **51**, 204 (1928)
Gurney, R. W. and Condon, E. U., *Phys. Rev.* **33**, 127 (1929)
von Laue, M., *Z. Phys.* **52**, 726 (1928)
Lit.: [3, 6d, 7, 9, 16, 17, 38 Vol. III, 49, 50 Vol. 4, part 3, 66, 70 Vol. II] (additional lit. see Table 20)

Two-way shape memory effect

→ memory effect

Tyndall effect (1869) Sc

(Tyndall phenomenon)

When light passes through a turbid solution, it is scattered at the molecules or particles. As the fluctuation in the number of molecules and the thermal motion of the molecules are independent one of the other, the scattered radiation is incoherent. As in → Rayleigh scattering, the scattered intensity is inversely proportional to the fourth power of the wavelength, so a greater proportion of short-wave light than long-wave light is scattered. The Tyndall-scattered light is thus bluer, whereas the light passing through the scattering body becomes steadily redder. Conclusions can be drawn on the anisotropy of the molecules in solution from the polarization (→ Rayleigh scattering, → Mie effect). The Tyndall effect resembles Rayleigh scattering in that it can be derived from the Mie theory of light scattering at spheres.

TYNDALL, John,
Irish physicist, 20 August 1820 – 4 December 1983, from 1853 Professor in London

Tyndall, J.: *Light: Six Lectures.*
Faraday, M., *Philos. Mag.* **14**, 512 (1857)
Tyndall, J., *Proc. R. Soc.* (London), **17**, 223 (1869)
Lord Rayleigh, J. W. Strutt, *Philos. Mag.* **41**, 107, 274, 447 (1871)

Lit.: [6c, 75] (additional lit. see Table 45)

Übler effect F

In the accelerated flow of a viscoelastic fluid through a narrowing tube, bubbles or suspended bodies may lag behind. The cause lies in the action of secondary normal stresses.

Lit.: *textbooks, dictionaries, handbooks*

Ueling effect (1935) AQ II

The deviation of the electrostatic potential of a point charge from the Coulomb potential as a consequence of vacuum polarization is called the Ueling effect.

UEHLING, Edwin Albrecht,
American physicist, 27 January 1901, Professor at the University of Washington

Lit.: [95, 100, 107, 114]

Undercooling effect S

(overheating effect)

The transition from the normal-conduction state to the superconducting state in a magnetic field is a first-order transition. Under certain conditions, a sample can be cooled below the critical temperature without superconduction setting in at once: the undercooling effect. The effect is fairly large for pure aluminum and indium.

The converse phenomenon, namely exceeding the critical temperature, is called the overheating effect.

Lit.: [11] (additional lit. see Table 44)

Uplift

→ Archimedes's principle

Valency effect S

By this is meant the shift in the transition temperature of a superconductor as a function of the concentration of doping atoms. These not only supply free electrons but also alter the lattice. The resulting internal mechanical stresses cause a marked dependence of the critical temperature on the doping atoms.

Lit.: [11] (additional lit. see Table 44)

Varistor effect SCC

Varistors are voltage-dependent resistors with symmetrical current-voltage characteristics, whose resistance decreases as the voltage increases. The exponential increase in the current with voltage which occurs when the threshold voltage is exceeded, is characterized by the nonlinearity constant. This has a value of about 5 for classical silicon carbide varistors, whereas for modern metal oxide varistors values lie between 20 and 30. The response time of a zinc oxide varistor is less than 25 ns, while the threshold voltage varies between a few volts and a few kilovolts.

The varistor effect is due to tunnelling through potential barriers at grain boundaries in the polycrystalline material.

A metal oxide varistor consists of zinc oxide, doped with other metal oxides, which has been sintered to produce a polycrystalline ceramic material. Varistors are used to protect components against excessive voltages in the entire field of electrical engineering and electronics.

Lit.: [91] (additional lit. see Table 49)

Vector effect

→ photoelectric effects, selective photoelectric effect

Villari effect (1865) Mm

In 1865, Villari discovered that a longitudinal tensile stress applied to iron or a ferromagnetic material generally increased the magnetization in weak fields, but reduced it in strong ones. The phenomenon is closely related to the extension of iron in the magnetization direction in weak fields and its contraction in strong ones. The origin of the effect is that an iron crystal shows positive magnetostriction in the tetragonal direction, but a larger negative magnetostriction in the direction of the body diagonals, so the → magnetostriction (Joule effect), taken as an average over all crystal directions, is negative. A material exhibiting positive magnetostriction increases in permeability in response to a longitudinal tensile stress, and vice versa.

VILLARI, Emilio,
Italian physicist, 25 September 1836 Naples – 20 August 1904 Naples
Villari, E., *Ann. Phys. Chem.* **126**, 87 (1865)
Lit.: [4, 26, 50 Vol. 4, part 4, 76] (additional lit. see Table 36)

Violet shift

→ gravitational frequency shift

Void effect N

In a sodium-cooled fast reactor, the reactivity is increased by overheating (boiling) the coolant. The resulting voids alter the structure of the material in electrical, mechanical, optical, and thermal respects.
Lit.: [90] (additional lit. see Table 38)

Voigt effect (1898) Mo

The Voigt effect is concerned with the *anomalous magnetic birefringence* derived from electron motion. It may be considered the counterpart of the → Faraday effect on transverse observation. The effect is very weak and can be observed only in the neighborhood of sharp strong lines in emission spectra, for example, the Na lines.
Voigt observed the effect on a sodium flame placed in a strong magnetic field. The effect is proportional to the square of the field strength.

VOIGT, Woldemar,
German physicist, 2 September 1850 – 13 December 1919, Professor in Königsberg and Göttingen
Voigt, W., *Nachr. Kgl. Gesellsch. d. Wissenschaften Göttingen* 355 (1898)
Voigt, W., *Wiedem. Ann.* **67**, 359 (1899)
Lit.: [22c, 50 Vol. II, 2.2, 75, 76] (additional lit. see Table 37)

Volta effect (1801) EM

A potential difference is set up when two different conductors are brought into contact (Schottky's term: *galvanic voltage*). The various metals can be arranged in a voltage series, which is given in the table below. The reason for this effect is that the electrons in a metal occupy all the energy levels up to the Fermi energy. If two different metals are brought into contact, electrons can pass by the tunnelling effect from the metal with the higher Fermi energy (and therefore the lower work function) to the other metal, but none in the reverse direction, since the energy levels are occupied. At the point of contact, the energy levels bend in order to match up.

Table 18. Volta's Voltage Series

+ Zn, Pb, Sn, Fe, Cu, Ag, Pt, C –

VOLTA, Alessandro Giuseppe Antonio Anastasio, Count
Italian physicist, 18 February 1745 – 5 March 1827, Professor in Como and Pavia
Volta, A., *Philos. Trans. R. Soc.* **90**, 403 (1800)
Volta, A., *Philos. Mag.* **7**, 289 (1800)
Volta, A., *Ann. Phys.* **10**, 421 (1802)
Volta, A., *Ann. Phys.* **12**, 497 (1803)
Volta, A., *Ann. Phys.* **14**, 257 (1803)
Volta, A.: *Collected papers of Count Alessandro Volta*, Comasco, Florence (1816)
Varney, R. N. and Fischer, L. H., *Phys. Unserer Zeit,* **9**, 107, 156 (1978)
Lit.: [50 Vol. 4, part 1, 55] (additional lit. see Table 26)

Voltage effect

→ Wien effect

Waterfall electricity

→ Lenard effect (1892)

Wave packet effect

→ diffraction filter-wave packet effect

Weissenberg effect Fl

In a non-Newtonian fluid normal stresses occur as a result of elasticity. If such a liquid is placed between two concentric cylinders, rotation of the inner cylinder causes the fluid to rise on the inner cylinder in spite of the centrifugal force.

WEISSENBERG, Karl,
Austrian physicist, 10 June 1893 Vienna, later (after 1933) England
Lit.: *[78], textbooks, dictionaries, handbooks*

Wertheim effect

→ Wiedemann effect

Wiedemann effect (1858) Mm

(circular magnetostriction effect)
If a ferromagnetic wire is placed in a longitudinal magnetic field and a current is passed through the wire producing a circular magnetic field in it, magnetization with a spiral distribution results: the circular magnetic field of the current and the longitudinal magnetic field are superimposed at each point in the sample.
Wiedemann showed that the wire twists as a result of magnetostriction. The converse is known as the *Wertheim effect* (1852).
When the sample is twisted in the longitudinal magnetic field, a potential difference appears between its ends.

WIEDEMANN, Gustav Heinrich,
German physicist, 2 October 1826 – 23 March 1899, from 1871 Professor in Leipzig
Wiedemann, G., *Ann. Phys. Chem.* **117**, 193 (1862)
Wiedemann, G., *Pogg. Ann.* **103**, 571 (1858)
Wiedemann, G., *Pogg. Ann.* **106**, 161 (1859)
Wertheim, A., *C. R.* **35**, 702 (1852)
Wertheim, A., *Ann. Chim. Phys.* **50**, 385 (1857)
Lit.: [6b, 26, 50 Vol. 4, part 4] (additional lit. see Table 36)

Wiegand effect EM

If a ferromagnetic wire is made from a magnetically soft core coated with a magnetically hard sheath, the system forms a bistable magnetic element. Appropriate mechanical and thermal stresses can cause the core and sheath to be magnetized in opposite directions. The core magnetization can be reversed in an external magnetic field. The sudden change in magnetization can be detected with an induction coil. There are two values for the external magnetic field at which the core magnetization in each case changes in direction.
At present, alloys are used containing, for example, 52 % cobalt, 38 % iron, and 10 % vanadium (vicalloy). With a wire diameter of 0.3 mm, a length of 30 mm, and an induction coil with 2600 turns, one gets a voltage amplitude of about 8 V.
The exact mechanism has not yet been finally established. The name *Wiegand wires* is given to wires with this property after their discoverer. Wiegand wire and coil can be used in a magnetic field sensor operating without a voltage supply and giving signals that are readily processed. However, the sensor can only detect the presence or absence of a magnetic field but not its magnitude.
Possible applications range from vehicle ignition, speedometers, and flow meters to security cards.
Dance, B., *Funkschau* **8**, 78 (1980)
Gevatter, H. J. and Meri, W. A., *rtp* **22, 3**, 81 (1980)
Kuers, G. and Waldhauer, G., *Electronik* **7**, 43 (1980)
Lit.: [100, 103, 114]

Wien effect (1927) El

(field-strength effect, voltage effect)
The electrical conductivity of a strong electrolyte increases steadily with the field strength at high values ($10^4 - 10^5$ V cm^{-1}). At very high field strengths the conductivity asymptotically approaches the limiting value for infinite dilution. As the ion drift velocity then attains a value of about 10 cm s^{-1}, the ion clouds no longer have time to form. Therefore, the → braking effects decrease. The ions finally move independently of one another.

WIEN, Max,
German physicist, 25 December 1866 – 24 February 1938, Professor in Aachen, Danzig, and Jena
Wien, M., *Ann. Phys.* (4) **83**, 327 (1928)
Wien, M., *Ann. Phys.* (4) **85**, 765 (1928)
Lit.: [50 Vol. 4, part 4] (additional lit. see Table 25)

Wigner effect N

When used in reactors, the physical properties of graphite are altered by fast neutrons. The lattice will take up large amounts of energy and is thereby disrupted. This energy can be released again on transition of the ground state.

WIGNER, Eugene Paul,
Hungarian-American physicist, 17 November 1902 Budapest, since 1930 Professor in Princeton

Lit.: [72] (additional lit. see Table 38)

Wijn aftereffect

→ magnetic aftereffects

Winslow effect

→ Johnson-Rahbeck effect

Zeeman effect (1896) Mo

In a weak magnetic field, a single spectral line splits into a triplet which may be explained in terms of Larmor electron precession. In the *transverse effect*, observation is made perpendicular to the magnetic field, and the undisplaced line is accompanied by new lines to its right and left, which exhibit right-handed and left-handed circular polarization.

In the *longitudinal effect*, the observation is made parallel to the magnetic field, and two lines with circular polarization are found symmetrically placed with respect to the original line.

The Zeeman effect occurring for a single line, for example for hydrogen, is called the *normal Zeeman effect*. In all multiple lines, the *anomalous Zeeman effect* is observed, in which the line splitting is substantially more complicated.

The Zeeman effect is used in astronomy, where the line splitting is used to obtain information on stellar magnetic fields. In strong fields, the Zeeman effect becomes the → Paschen-Back effect.

In solids, the Zeeman effect as well as the Stark effect can be observed under certain conditions.

→ Stark effect

ZEEMAN, Pieter
Dutch physicist, 25 May 1865 – 9 October 1943, from 1900 Professor in Amsterdam

Zeeman, P., *K. Ned. Akad. Wet. Versl. Gewone Vergad. Afd. Natururkd.* **6**, 13, 99, 260 (1897)
Sommerfeld, A., *Phys. Z.* **17**, 491 (1916)
Debye, P. *Phys. Z.* **17**, 512 (1916)

Lit.: [3, 6c, 6d, 17, 22c, 27, 41, 50 Vol. II, 2.2, 66, 69, 70 Vol. II, 90, 93] (additional lit. see Table 37)

Zeeman effect, inverse Mo

If an absorbing medium is placed in a magnetic field, the absorption lines exhibit the Zeeman effect, which corresponds exactly to the emission effect.

A consequence of the inverse Zeeman effect is that there is anomalously large magnetic birefringence in the neighborhood of an absorption line.

The inverse Zeeman effect is observed particularly in the rare earths (cerium, lanthanum, neodymium, praseodymium, etc.). The line splitting in an anisotropic crystal can be very complicated.

Lit.: [50 Vol. II, 2.2] (additional lit. see Table 37)

Zener effect (1934) SCC

If the bias voltage applied to a p-n junction is increased, the conduction and valency bands may switch over, the consequence being that the → tunnel effect leads to electrons passing from the valency band into the conduction band. A rapid increase in the current is then observed. The Zener effect is used in Zener diodes for voltage stabilization.

At higher voltages, the Zener breakdown gives way to avalanche breakdown. The Zener effect is one of the → high-field effects characteristic of semiconductors.

→ avalanche effect, → Gunn effect

Zener, C., *Proc. R. Soc. London Ser. A* **145**, 523 (1934)

Lit.: [83, 84, 85, 91, 92] (additional lit. see Table 49)

Appendix

Peripheral Areas and Special Effects

Cosmic-radiation effects

Azimuthal effects
On account of the Earth's magnetic field, the distribution and intensity of cosmic radiation at the Earth are dependent on azimuth, although they are isotropic in space (*east-west asymmetry*).

Barometer effect
The absorption of cosmic radiation in the atmosphere varies with the atmospheric pressure at the point of observation.

Blanketing effect
The Fraunhofer lines absorb 9 % of the radiation flux in the photosphere, but about half is re-emitted into the photosphere and produces a temperature rise in the outermost layers. When allowance is made for this effect, which is a form of *glasshouse effect*, there is very good agreement between observation and calculation for the margin darkening of the photosphere on the Sun's disc.

Glasshouse effect
(hothouse effect, greenhouse effect)
This occurs because sunlight passes through glass without substantial loss, whereas the glass absorbs the longer-wavelength thermal radiation. In a building consisting mainly of glass, such as a hothouse, sunlight is admitted, and by absorption in the interior is converted to heat, which increases the temperature.
The same effect occurs in the Earth's atmosphere, where the strong absorption by atmospheric water vapor and carbon dioxide for thermal radiation prevents the heat from being radiated away from the Earth's surface. On account of the considerable production of carbon dioxide, this effect is of great significance for the sensitive thermal equilibrium on the Earth.
The dense atmosphere of Venus exhibits the same effect, which leads to the high temperatures on that planet.

Latitude effect
The cosmic-radiation intensity increases with geomagnetic latitude, because at high latitudes the Earth's magnetic field has a greater effect on charged particles.

Longitude effect
The longitude effect is produced by the Earth's magnetic field, which is a dipole field whose axis does not coincide with the Earth's axis and which also does not pass through the center of the Earth. The effects of the Earth's magnetic field on cosmic radiation are therefore dependent on the geographical longitude.

North-south effect
There is a slight difference between the intensities of cosmic radiation incident from the north and south. In the Northern hemisphere, the intensity incident from the south is somewhat greater than that from the north, and vice versa. The effect is explained in terms of geomagnetic effects, e. g., the particle paths being influenced by the Earth's magnetic field, and especially by the Earth's shadowing action.

Transition effect in cosmic radiation
Transition effect is the name given to the fact that the radiation absorption alters when there is any change in the absorbing material. In spite of conversion to equivalent layers, the absorption differs to that observed had the material remained the same.

Twilight effect
During morning and evening twilight, the spectrum shows accentuation of the yellow sodium lines, the red oxygen doublet, and a somewhat smaller increase in the emission of the positive nitrogen band.
The explanation is to be sought either in photodissociation produced by ultraviolet sunlight or in optical resonance between the sodium atoms in the yellow radiation from the Sun. A similar explanation applies for the oxygen and nitrogen lines. The effect occurs at heights between 80 and 100 km.

Development effects
(see also Photographic effects)

Lainer effect
The addition of potassium iodide to the developer reduces the induction period and thus accelerates the development, the reason for this

being that potassium iodide tends to dissolve silver bromide.

Neighbor effect

During the development of a photographic layer, the blackening can be affected by blackening in the neighborhood if the developer is not agitated (also *Eberhard effect*). The diffusion of developer from slightly exposed areas to heavily exposed areas produces a change in the photographic pattern, with, for example, sharper edges (the *edge effect, border effect, margin effect*), or else undesired borders. The modulation transfer function of the layer is improved (*Kostinsky effect*).

Ross effect

(gelatin effect)

On drying of a photographic layer, blackened areas tend to shrink. Allowance must be made for this effect when evaluating plates differing very greatly in exposure, because e.g. blackened circles become smaller in diameter or the distances between lines tend to increase.
The effect can be reduced by alcohol drying.

Temperature effect

The sensitivity of a silver halide decreases at low temperatures.

Illumination effects

Albert effect

The Albert effect resembles the Debot effect. If the development centers on the surface of a previously exposed photographic film are destroyed by the use of chromic acid, image inversion can be produced by fresh exposure.

Becquerel effect

Diffuse illumination causes unsensitized silver halide to become sensitive in the long-wave region. The contrast can be increased by a factor 4 or 5, for example, in X-ray pictures.

Cabannes-Hofmann effect

If a photographic film is exposed for a long time at a low intensity, during development blackening occurs more rapidly than if the same amount of light had been provided in a shorter time at a higher intensity. The final blackening density is the same in the two cases.

Clayden effect

If a photographic film that has been exposed to a brief flash is further illuminated, there is often a reduction in density (black lightning in photographs of thunderstorms). The reversal of an X-ray picture by normal light is called the Villard effect.

Debot effect

If the surface nuclei in an exposed photographic emulsion are removed with chromic acid, the picture can be restored by exposure to infrared light, since nuclei are thereby transferred from the interior of the layer to the surface.

Herschel effect

The latent picture in a photographic film which is normally not red-sensitive can be partially destroyed by exposure to red light.

Intermittency effect

Talbot's law states that the eye has the same impression of brightness from light stimuli separated by dark spaces as when the same total light stimulus is provided uniformly distributed at a lower intensity. In a photographic film, there is a reduction in density when the number of interruptions per second is very large (*intermittency effect*).

Schwarzschild effect

This is the deviation between the density of a photographic film and that predicted by the reciprocity law on long exposure and at high light intensities.

Weigert effect

This effect, known since 1919, may be due to photoelectric phenomena and is worthy of note because it represents a rare case of a special effect due to linearly polarized light. If a blue-red layer, or photochloride (i. e., an adsorption compound of silver chloride with metallic silver, such as occurs when gelatin loaded with silver chloride is left exposed to light for a certain time), is exposed to high-intensity linearly polarized light, the blue-red film becomes dichroic, and the distinctive direction coincides with the direction of the electric vector. It is thus possible to demonstrate photographically the plane of polarization of the light.

Weiland effect
If a photographic material is exposed for a short time at a very high intensity and for a long time at a lower intensity, a higher degree of blackening results than with the converse sequence. The brief intense illumination produces tiny development centers which then grow to full ones.

Perception effects

Aubert phenomenon
An observer in a darkened room sees a perpendicular line as oblique if he inclines his head to one side. The head inclination and the oblique setting are in general in opposite directions.
This is called the *E phenomenon* when the directions are the same.

Bezold-Abney phenomenon, Brücke-Abney phenomenon
At very low intensities the eye can distinguish only three colors: blue-violet (380 nm – 480 nm), green (480 nm – 570 nm), and red (570 nm – 760 nm). The fact that no other colors can be distinguished is called the *Brücke-Abney phenomenon*.
As the light intensity increases, all colors can be distinguished (160 tones and 30 purple colors). Any further increase produces no further difference. If the light is perceived as colorless at very high intensities, one speaks of the *Bezold-Abney phenomenon*.

Emmert's phenomenon
The visual size of afterimages changes when the distance between the projection surface and the eye alters. The change in visual size corresponds to the change in the retinal image.

Geld-Benussi phenomenon
(τ phenomenon)
If an observer is presented consecutively with three spatially equidistant light sources lying on a line, the eye matches the spatial and time separations: the observer sees the light stimuli at other points and at other times.

Herzau-Ogle phenomenon
If there is unequal enlargement in the two eyes, or if corresponding retinal positions do not correspond to the same geometrical points, identical figures seen with stereoscopic vision may appear as different and therefore double. In the same way, figures of the same shape but different in size can appear as identical. This fact is known as the Herzau-Ogle phenomenon.

Purkinje effect
By this is meant the shift in the relative brightness of colored lights on changing between daylight and twilight vision.

Stiles-Crawford effect
(aperture effect)
In various photometric methods, it is important to make allowance for locally variable transmission in the eye lens (the *Stiles-Crawford effect*). The light source and the comparison source must be imaged on the same part of the pupil and be equally large.
If the peripheral areas of the pupil are used, there may be an impairment of color perception (*Stiles-Crawford effect of the second kind*).

Photoelectric dark effect

Dark currents produced in photocells by leakage currents, barrier-layer currents, and thermal effects will give rise to interference.

Photographic effects

(see also Development effects)

Callier effect
Diffuse and parallel light expose a photographic layer to different extents. The blackening is greater for parallel light. The name *Callier coefficient* is given to the parallel/diffuse blackening ratio.

Sabattier effect
If an exposed photographic layer is reexposed during development, reversal of the first picture is obtained. The areas already blackened act as shadows in the second exposure.

Radio-wave effects

High-light effect
This is an effect in the propagation of electromagnetic waves. When waves of the same frequency but from different directions arrive at a direction-finding aerial with a phase difference of 180°, an error of almost 90° can occur.

Luxembourg effect
The modulation of a weak transmitter can be influenced by strong nonlinear effects in the ionosphere, occurring during reception (at long and medium wavelengths, as is observed with Radio Luxembourg which can be heard at the Beromünster frequency).

Mögel-Dellinger effect
(solar glare effect)
Short-wave telecommunications are interrupted during solar glares, in which the lower ionosphere is ionized by UV radiation.

Night effect
a) The polarization of the space wave carrying medium-wave and long-wave reception at night undergoes persistent or temporary changes.
b) Effect of a) on the direction-finding error: In the early days of direction finding, the night effect was not familiar, but nowadays its fateful consequences are restricted by using methods free of the night effect.

PCA effect
(polar cap absorption)
In strong solar flares extremely heavy ionization can lead to very strong damping of radio waves over the polar regions. This effect can persist for several hours or days and results in a breakdown of radio communications in these regions.

Solar flare effects
This is a characteristic disturbance of the Earth's magnetic field in which the declination alters by up to 10 angular minutes within a few minutes of time, while the horizontal intensity alters by up to 50 γ, finally slowly returning to their normal values. The effect often occurs at the same time as the Mögele-Dellinger effect. An explanation has been given in terms of currents in the D layer.

Thunderstorm effect
(asymmetry effect)
During a thunderstorm, the electric field strength increases, which produces interference in the frequency range from 100 to 3500 MHz.

Twilight effect
(night effect)
The height of the ionosphere alters during twilight and in the night, which interferes with radio communications.

Tube effects

Clean-up effect
(gettering)
The deposition of ions on the walls or other parts of a device in which ions are produced in a vacuum is called the clean-up effect or gettering. The effect is used in producing a high vacuum: in the production of vacuum tubes pumping times are reduced and in sealed tubes the vacuum is maintained in spite of slight gas release from the electrodes by means of the property of various substances of taking up gases in substantial amounts by adsorption, absorption, or chemical bonding. Substances with such properties are called getters and the process is known as gas cleanup or gettering. In addition to getters, cooled surfaces and drying agents are used, especially for the removal of water vapor.

Dark effect
This is a interfering effect in photocells, occurring on the measurement of very low light intensities and due to thermal electron motion.

Dynatron effect
A current-voltage characteristic with negative slope may be obtained due to secondary emission from an electrode at a lower potential than a preceding electrode in the line of motion of the primary electrons. The name dynatron is given to an electron tube that employs this effect to produce oscillations without coupling. The dynatron no longer has any technical significance.

Emission-drift effect in electron tubes
Emission changes with frequencies of less than 0.1 Hz can occur for electrons produced by oxide cathodes in tubes.

Flicker effect
Individual surface elements in a photocathode are displaced spontaneously to a greater or lesser extent, so that the mean total surface emission also changes spontaneously. This becomes apparent as a rustling and crackling noise. The flicker effect should be distinguished from the shot effect. The Funkel effect in thermionic cathodes is related to the flicker effect.

Malter effect
This is a charging phenomenon occurring in counting tubes containing light-metal cathodes. The effect produces a delayed discharge which results in spurious pulses, so the count rate is falsified.

Microphone effect
Mechanical vibrations of the components in a valve can modulate the electron current. Acoustic feedback, e.g., from a loudspeaker, can result in self-excitation, which is observed as a whistle.

Miller effect
If a capacitor is connected between grid and anode, a voltage decrease occurs at the anode if the grid voltage rises. If the voltage between grid and cathode is altered, the capacitor receives a voltage swing $(1 + V)$times larger (V is the amplification of the stage).
The Miller effect itself does not influence the linearity of the voltage rise. Only small capacities should be used in order to obtain suitable voltage amplitudes.

Pätow effect
In counting tubes filled with noble gases delayed discharges occur because metastable excited states release secondary electrons. These produce spurious pulses.

S effect
Malfunctions in power pentodes for high alternating anode voltages, which are caused by the sudden charging of insulators in these electron tubes on account of collision by primary electrons, for example, at the inner surfaces of the glass bulb.

Space-charge effect
The origin of this effect lies in the negative electron charge. Repulsive forces between electrons affect the motion in a flow of electrons. The space charge causes divergence in focussed electron beams. Further, it affects the spatial distribution of an external electric field (Poisson's equation). Space charge is also responsible for the occurrence of a virtual cathode in an electron tube.

Stoletov effect
The photocurrent in a gas-filled photocell is dependent on the gas pressure at constant voltage and constant illumination. On admitting an inert gas in steps to a vacuum photocell, the photocurrent increases, and the number of ionizing collisions of electrons with atoms increases with the pressure. When a certain pressure is exceeded, the photocurrent decreases again, because ionization is reduced due to the decreasing electron mean free path.

Tables

Table 19. General effects (G)

Cause	Effect	Name
Electric and/or magnetic fields acting on materials	Production of electric or magnetic moments in field direction	Alignment effects
Different structures in different directions	Direction dependence of effects	Anisotropy effects
Anisotropic optical materials	Production of waves with mutually perpendicular polarization	Birefringence
Various causes, e. g., electric fields	In the bulk of a body, e. g., production of domains	Bulk effects, e. g., Gunn effect
Slits or screens in the path of rays (in general, a wave)	Deviation from straight-line propagation	Diffraction
Internal nonequilibrium distribution	Particle transport	Diffusion
Wavelength dependence of refractive index	Change in phase velocity, frequency dependence of material constants	Dispersion
Overlap between coherent waves of the same wavelength	Wave reinforcement or attenuation	Interference
External forces/ fields acting on bodies	Time-delayed start of actions	Relaxation effects Aftereffects
Atomic or subatomic particles or their fields in the paths of waves or particles	Deflection of waves or particles from propagation direction	Scattering effects, scattering of electromagnetic waves, scattering of elementary particles, scattering of charge carriers in solids
Pulsed light on moving bodies	Apparently motionless state in moving bodies	Stroboscope effect
Unequal wave or particle transit times	Changes in pulse shape	Transit-time effects
Irreversible motion of particles under external forces	Various, e. g., viscosity, thermal conduction, diffusion, conductivity, cross effects	Transport effects

Lit.: *textbooks, dictionaries, handbooks, additional literature, cited on pages 135 ff.*

Table 20. Atomic and quantum physics I (AQ I)

Cause	Effect	Name
Nuclear motion	Change in electron binding energy	Comovement effect
Different isotopes	Spectrum change	Coupling effect
Interaction between radiating gas atoms	Level shift	Coupling effect
Force between molecular dipoles	Contribution to van de Waals forces	Dipole effect (Richt effect)
Electron motion	Fluctuating atomic or molecular dipoles	Dispersion effect
Exchange of indistinguishable particles	Energy shifts of spectral lines, saturation of chemical bonds, etc.	Exchange effects
Force between molecular dipoles at high temperatures	Additional electric moments	Induction effect
Effect of nuclear potential on gamma radiation near nucleus	Production of electron-positron pairs	Pair production effect
Electron penetration into atomic core	Change in electron binding energy	Penetration effect
Perturbation in the electron distribution in a molecular band, e. g., produced by external fields	Perturbation propagation through molecular bands	Substituent effect (perturbation propagation effect, I or induction effect, F or field effect, mesomeric effect)
Motion of a particle in a potential well	Tunnelling through potential barrier	Tunnelling effect

Lit.: [3, 6d, 7, 17, 22d, 38 Vol. III, 41, 46, 60, 66, 70 Vol. II, 79, 90, 95, 96, 100, 101, 107, 111, 114, 118, 121]

Table 21. Atomic and quantum physics II (AQ II)

Cause	Effect	Name
Internal absorption of radiation in atomic shells or in solids	Occurrence of characteristic electrons	Auger effect, internal photoionization, internal photoelectric effects
Retardation of potential between neutral atoms, effects of zero-point energy	Attractive force between plates in vacuum	Casimir effect
Passage of fast charged particles through a dielectric medium	Occurrence of a characteristic radiation cone in the blue region of the spectrum	Cherenkov effect

Table 21 (continued)

Cause	Effect	Name
X-ray scattering at electrons in matter	Increase in wavelength of scattered radiation	Compton effect
Particle production	Simulation of a resonance	Deck effect
Absorption of circularly polarized light by an alkali-atom beam	Production of spinpolarized electrons	Fano effect
Nonvanishing charge distribution in neutron	Interaction of the electric field in the neutron with shell electrons	Foldy effect
Optical transitions in atoms or molecules	Virtually no change in nuclear separation and in momentum of atoms or molecules	Franck-Condon principle
Collisional excitation of mercury atoms by electrons in a tube	Characteristic current-voltage curves with steps at multiples of the excitation energy	Franck-Hertz experiment
Conduction-electron paramagnetism	Shift in nuclear resonance frequency to higher values	Knight shift
Electron vacuum polarization	Spectral line shift	Lamb shift (Bethe effect)
Collision of neutral atoms with hot metal surfaces under vacuum	Production of ions from atoms	Langmuir effect
Collision of low-energy photons with gamma rays	Production of electron-positron pairs	Nikishov effect
Production of a neutral π meson	Decay of the π_0 into two γ rays	Primakoff effect
Scattering of slow electrons at inert-gas atoms	Reduction in interaction cross section	Ramsauer effect
Action of laser light on aromatic molecules at low temperatures	Characteristic narrow fluorescence lines	Shpol'skii effect
Electron beam striking a metal grating	Production of coherent radiation	Smith-Purcell-Salisbury effect
Inhomogeneous magnetic field acting on an atomic beam	Atomic beam splitting	Stern-Gerlach experiment (Stern-Gerlach effect)
Vacuum polarization	Deviation of electrostatic potential of a point charge from the Coulomb potential	Uehling effect

Lit.: [3, 6d, 7, 17, 22d, 41, 46, 60, 66, 70 Vol. II, 78, 79, 90, 95, 96, 100, 101, 107, 111, 114, 118, 121]

Table 22. Astronomy (As)

Cause	Effect	Name
Red shift	Decrease in light frequency and therefore in photon energy	Energy effect
Stellar recession	Red shift in spectral lines increases with the distance	Hubble effect
Astronomical objects receding from observer	Lower brightness	Number of dilution effect
Expansion of the universe	Stellar systems receding from us	Stellar recession

Lit.: [10, 14, 50 Vol. 5, part 2, 80, 100, 111, 112, 114, 118a, 120]

Table 23. Current conduction (CC)

Cause	Effect	Name
Current-carrying conductor in a transverse or longitudinal magnetic field	Potential and/or temperature differences	Galvanomagnetic effects
Current passing through a homogeneous conductor with an inhomogeneous temperature distribution	Temperature difference, thermo-EMF	Thermoelectric homogeneous effects
Current passing through two different conductors connected together	Temperature difference, thermo-EMF	Thermoelectric inhomogeneous effects
Heat flux passing through a conductor in a transverse or longitudinal magnetic field	Potential and/or temperature differences	Thermomagnetic effects

Lit.: [1, 3, 6b, 20, 22b, 23, 30, 38 Vol. VIII, 43, 49, 50 Vol. 4, part 4, 53, 55, 67, 86 Vol. 13, 89 Vol. 16, part 1, 94, 96, 100, 101, 104, 110, 114]

Table 24. Electrokinetics (Ek)

Cause	Effect	Name
Action of the gravitational potential on charged particles in a liquid	Potential difference between surface and bottom	Dorn effect
Electric field acting on a liquid	Change in surface tension	Electrocapillarity
Electric field acting on charged particles in a liquid	Motion of charged particles through a diaphragm	Electroosmosis, electroendosmosis
Electric field acting on charged particles in a liquid	Charged particle drift in a fluid	Electrophoresis, cataphoresis, anaphoresis
Addition of electrolytes to a solution	Viscosity change	Electroviscous effect
Fluid flow in a capillary	Flow potential, apparent viscosity increase	Electroviscous effect
Liquid flowing through a diaphragm	Flow stresses on both sides of diaphragm	Flow current, flow potential

Lit.: [6b, 6d, 12, 23, 24, 29, 100, 113, 114]

Table 25. Electrolytes (El)

Cause	Effect	Name
Voltage applied to an electrolyte	Asymmetric ion-cloud distribution around an ion, drift inhibition	Braking effect (drift inhibition, asymmetry or relaxation effect)
Molecular dissociation	Diffusion inhibited by other molecules, back-reaction	Cage effect
High-frequency alternating field in an electrolyte	Conductivity increase	Debye-Falkenhagen effect (conductivity dispersion effect)
Strong electric field applied to a weak electrolyte	Change in dissociation, increase in conductivity	Dissociation-voltage effect
High voltage applied to a strong electrolyte	Increase in electrical conductivity	Wien effect (field-strength or voltage effect)

Lit.: [6b, 6d, 12, 22d, 24, 50 Vol. 4, part 4, 100, 114]

Table 26. Electricity and magnetism (EM)

Cause	Effect	Name
The surface layer involved in conduction at high frequencies and low temperatures is small compared with the mean free path	Conductivity independent of mean free path	Anomalous skin effect
Rotation of a copper disc	Drag on magnetic needle	Arago's experiment
Magnetization reversal in iron	Switching in elementary domains	Barkhausen effects
Addition of rare earths to carbon anode	Light intensity increase	Beck effect (Beck arc)
Perpendicular magnetic fields in thin orthoferrite layers	Production of cylindrical magnetic domains, which are mobile in the layer	Bobeck effect, magnetic memory effect
Electric field acting on dielectric	Displacement of associated ions	Böning effect
Small magnetic flux or vector potential	Change in electron beam interference	Bohm-Aharonov effect
High electric fields at points	Brush discharge	Corona effect, St. Elmo's or Elias fire
Electric field	Alignment of the permanent dipoles in atoms or molecules	Debye effects
Residual magnetism in the iron of a stator coil carrying current in a generator	Magnetic-field self-excitation up to saturation	Dynamoelectric principle
Polarization reversal in a ferroelectric	Switching in ferroelectric domains	Ferroelectric Barkhausen effect
High voltage acting on fine metal points under high vacuum	Emission of electrons	Field effect
Rotation or magnetization of an iron rod	Magnetization or rotation of rods	Gyromagnetic effects (Barnett effect, Einstein-de Haas effect)
Passage of linearly polarized electromagnetic waves through metal rods	Radiation attenuation in accordance with rod setting	Hertz's experiments, Hertz effect
Weak magnetic fields acting on a ferromagnetic a little below the Curie temperature	Peak permeability in ferromagnetic	Hopkinson effect

Table 26 (continued)

Cause	Effect	Name
Polarization of permalloy or mu metal	Asymmetries in alternating-current hysteresis	Hughes effect
Current passing through a metallic conductor	Conductor heating	Joule effect
High static voltage on an insulating body	Discharge figures	Kirlian effect, Lichtenberg figures
Removal of surface layers or water droplets	Charging of droplets and water	Lenard effect, Ballo-electricity, spray effect, waterfall electricity
Magnetic or electric field applied to an insulating material with magnetic ordering	Proportional electric polarization and proportional magnetization	Magnetoelectric effect
Metal surface heating	Electron emission	Richardson effect (Edison effect, thermionic effect)
Rotation of dielectric in a charge condenser	Röntgen current	Röntgen-Eichenwald experiment
Rotation of a charged condenser plate	Convection currents, Rowland current	Rowland effect
Displacement of eddy currents in a conductor at high frequencies	Increase in resistance	Skin effect
Contact between two different conductors	Production of contact voltage	Volta effect
Magnetization reversal in a coated ferromagnetic wire	Detection of magnetization reversal by an induction coil	Wiegand effect

Lit.: [3, 6b, 16, 18, 22b, 26, 30, 32, 37,38 Vol. VIII, 40, 43, 49, 50 Vol. 4, 69, 100, 101, 103, 105, 109, 111, 112, 114, 115]

Table 27. Electrooptics (Eo)

Cause	Effect	Name
Electric field acting on an optically isotropic medium	Optical anisotropy, birefringence	Kerr effect, electrooptic birefringence
Electric field acting on an optically isotropic medium	Change in polarization	Neugebauer effect
Electric field acting on a piezoelectric crystal	Birefringence	Pockels effect, linear electro-optic effect
High electric field gradients at the poles of an electric arc	Slight shifts in wavelengths near poles relative to middle of arc	Pole effect
Electric field acting on atoms or molecules emitting light or radiation	Spectral line splitting	Stark effect

Lit.: [3, 6c, 6d, 9, 17, 22c, 50 Vol. II, 2.2]

Table 28. Electromechanics and electrothermics (ET)

Cause	Effect	Name
Electric field acting on a crystal without a center of symmetry	Temperature change in crystal	Electrocaloric effect
Electric field acting on an insulator	Deformation or volume change	Electrostriction
Mechanical stress on crystals	Entropy production	Piezocaloric effect (converse: thermal expansion)
Mechanical stress on a crystal without a center of symmetry	Electrical potential, change in polarization	Piezoelectric effect, piezo effect (direct)
Electric field acting on a crystal without a center of symmetry	Mechanical deformation	Piezoelectric effect, piezo effect (inverse)
Mechanical stress on a metal or semiconductor	Change in electrical resistance	Piezoresistance effect
Rapid temperature change in a crystal without a center of symmetry	Surface charging, electrical potential	Pyroelectric effect

Lit.: [6b, 18, 27, 38 Vol. VIII, 50 Vol. 4, part 4, 94, 95, 96, 100, 102, 103, 109, 114]

Table 29. Flow (F)

Cause	Effect	Name
Liquid flow	Deviation from flow direction	Coanda effect
Gas flow	Change in up thrust with height	Ground effect
Moving projectile with spin	Force acting on rotating body	Magnus effect
Liquid flow	Expansion on emergence from a nozzle	Merrington effect
Moving projectile with spin	Air cushion	Poisson effect
Turbulent gas flow in a tube	Change in gas temperature	Ranque effect
Accelerated flow of a visco-elastic fluid through a narrowing tube	Bubbles or suspended bodies remain stationary on account of normal stresses	Übler effect
Liquid flow	Change in normal stress	Weissenberg effect

Lit.: *textbooks, dictionaries, handbooks*

Table 30. Galvanomagnetism (Ga)

Cause	Effect	Name
Conduction-electron scattering and magnetic moments in thin ferromagnetic films	Anomalously large Hall constant	Anomalous Hall effect
Transverse magnetic field in a current-bearing conductor	Transverse temperature difference	Ettingshausen effect
Same cause as in the von Klitzing effect	Establishment of a multi-particle ground state	Fractional quantum Hall effect
Conductor carrying in a transverse magnetic field	Transverse potential difference	Hall effect, Corbino effect (dissymmetry effect)
Current-carrying conductor in a longitudinal magnetic field	Resistance change	Magnetoresistance effect
Current-bearing conductor in a transverse magnetic field	Longitudinal temperature difference	Nernst effect
Transverse magnetic field in a current-bearing conductor	Longitudinal potential difference	Thomson effect, magnetoresistance effect
Current-carrying thin semiconductor films (MOS transistors) or semiconductor inversion films in strong transverse magnetic fields at low temperatures	Electron-orbit or Hall-voltage and Hall-resistance quantization	Von Klitzing effect
Current-carrying conductor partially in a longitudinal magnetic field	Peltier effect between magnetic and nonmagnetic conductors	(No name)

Lit.: [1, 3, 6b, 20, 22d, 23, 30, 38 Vol. VIII, 43, 49, 50 Vol. 4, 4, 53, 62, 67, 94, 100, 101, 104, 107, 110, 114]

Table 31. Lasers and nonlinear optics (L)

Cause	Effect	Name
Coherent light incident on an object, reflection from the surface	Interference pattern of the object surface	Holography
Intense laser light in liquids	Shock waves	Light-hydraulic effect
Nonlinear relationship between polarization and electric field	AC Kerr effect, Bragg effect, Frequency doubling, hyper-Raman effect, lens effect, optical mixing, optical rectification, optical Kerr effect, Raman-induced Kerr effect, self-focussing, induced scattering, gas ionization	Nonlinear optical effects
Excitation of atoms or molecules by, e.g., light	Population inversion	Stimulated radiation emission, laser effect

Lit.: [3, 6c, 19, 22c, 32, 33, 42, 47, 48, 56, 57, 67, 96, 98, 100, 106]

Table 32. Liquid crystals (LC)

Cause	Effect	Name
Electric field acting on a liquid crystal	Current flow and mechanical liquid flow	Electrohydrodynamic effect
Electric field acting on a thin liquid crystal layer	Rotation of molecular axes	Field effects: DAP effect, guest-host effect, Schadt-Helfrich effect
Electric field acting on a thin liquid crystal layer	Switch between different liquid-crystal textures	Memory effects, texture-conversion or bistability effect

Lit.: [6d, 100, 114]

Table 33. Low temperatures (LT)

Cause	Effect	Name
Sudden major change in an external electric field on a sample	Temperature reduction of a few degrees due to additional degrees of freedom	Electrocaloric effect
Sudden strong magnetization of a paramagnetic salt	Salt temperature reduction	Magnetocaloric effect
Quantum properties of He II	Cold transport	Onnes effect (mechanocaloric effect, thermodynamic pressure effect, fountain effect)
Adiabatic compression of liquid ^3He	Decrease of temperature	Pomeranchuk effect
Cooper pair production at low temperatures	Loss of resistance, occurrence of persistent current, Meissner effect, etc.	Superconductivity (see Dictionary of Effects and Table 41 (Cooper effect)

Lit.: [2, 6d, 11, 15, 22d, 29, 38 Vol. V, Vol. IX, 41, 50 Vol. II, part 4, 100, 114]

Table 34. Luminescence (Lu)

Cause	Effect	Name
Direct contacts on high-conductivity crystals	Field quenching	Déchêne effect
Alternating voltage on phosphors	Emission from phosphors	Destriau effect
Electric field on a crystal	Increase in light emission	Gudden-Pohl effect
Current through a SiC rectifier	Difference in luminescence at cathode and anode	Lossew effect
Increased crystal stimulation	Increased thermoluminescence	Riehl effect

Lit.: [6d, 75, 78, 84, 96, 97, 100, 101, 114]

Table 35. Mechanics (acoustics, motion, heat) (Me)

Cause	Effect	Name
Hydrostatic pressure on a body in a liquid	Occurrence of a force acting in opposition to gravity	Archimedes' principle
Movement of head in hearing	Location of sound source	Binaural effect
Rotating reference system	Additonal force	Coriolis force
Motion of a sound source	Frequency change	Doppler effect
Motion of a sound source	Change in terrestrial acceleration	Eötvös effect
Motion of a sound source	Rotation of the plane of a pendulum	Foucault's experiment
Two sound sources with a phase difference	Sound source masking	Haas effect
Liquid droplets on a hot surface	Insulating vapor layer	Leidenfrost phenomenon
Sound waves in a wall	Bending movement in the wall	Line adaptation effect and coincidence effect
Two closely spaced frequencies	Masking of the higher frequency	Masking effect

Lit.: *textbooks, dictionaries, handbooks*

Table 36. Magnetomechanics (Mm)

Cause	Effect	Name
Magnetization of a ferrimagnetic or ferromagnetic sample	Elasticity change	ΔE effect, magnetoelastic effect
Stresses in a ferromagnetic sample	Change in remanence	Magnetic tension effect
Magnetization of a ferrimagnetic or ferromagnetic sample	Volume or shape change	Magnetostriction, Joule effect
Torsion of a magnetized rod in a longitudinal magnetic field	Magnetization change	Matteucci effect
Alternating magnetic field of small amplitude acting on a magnetized sample	Shape change linearly proportional to the field	Piezomagnetic effect

(continued on page 112)

Table 36 (continued)

Cause	Effect	Name
Mechanical oscillations in a magnetized sample	Magnetization change proportional to the rotation	Piezomagnetic effect
Stresses in a ferromagnetic sample	Permeability change on sample contraction or extension	Villari effect
Torsion of a magnetized rod in a longitudinal magnetic field	Induction currents produced	Wertheim effect
Current along the axis of a magnet or rod in a longitudinal magnetic field	Rod torsion	Wiedemann effect

Lit.: [4, 6b, 26, 50 Vol. I4, part 4, 100, 114]

Table 37. Magnetooptics (Mo)

Cause	Effect	Name
Strong magnetic field acting on atoms or molecules	Hyperfine interaction between shell and nucleus	Back-Goudsmith effect
Longitudinal magnetic field acting on a paramagnetic substance	Rotation of plane of polarization	Becquerel effect, paramagnetic rotation
Longitudinal magnetic field acting on an optically isotropic medium	Rotation of plane of polarization	Faraday effect (magnetorotation), magnetic rotation of the plane of polarization
Magnetic field acting on Hg vapor also exposed to polarized light	Decrease in polarization of resonant radiation	Hanle effect
Magnetic field acting on a ferromagnetic film	Anomalously large rotation of plane of polarization	Kundt effect
Longitudinal magnetic field acting on a gas	Rotation of the plane of polarization near absorption lines	Macaluso-Corbino effect
Weak magnetic fields acting on atoms or molecules	Rotation of plane of polarization in fluorescent light	Magnetic rotation effect (Hanle effect)
Reflection of linearly polarized light from the polished pole faces of a magnet	Rotation of plane of polarization of reflected light	Magnetooptic Kerr effect
Transverse magnetic field acting on a colloidal solution	Birefringence	Majorana effect

Table 37 (continued)

Cause	Effect	Name
Strong magnetic field acting on light-emitting atoms or molecules	Elimination of coupling between spin and orbital angular momentum	Paschen-Back effect
Magnetic field perpendicular to light propagation direction in a liquid containing anisotropic molecules	Birefringence	Transverse magnetic birefringence: Cotton-Mouton effect
Transverse magnetic field acting on a gas, e.g., Na vapor	Anomalous birefringence for strong spectral lines	Voigt effect
Magnetic field acting on light-emitting atoms or molecules	Line splitting	Zeeman effect, inverse Zeeman effect

Lit.: [3, 6c, 9, 18, 22c, 48, 50 Vol. II, 2.2, 57, 67, 110, 114]

Table 38. Nuclear physics (N)

Cause	Effect	Name
Unequal charge number, equal mass number, equal isospin, unequal third component	Energy difference	Analogue states, Coulomb effect, mass effect
Incorporation of nuclei into crystal at low temperatures	Recoil-free gamma-ray absorption and emission	Mössbauer effect
External magnetic field acting on excited nuclei in solids	Radiation depolarization	Nuclear Hanle effect
Electromagnetic waves acting on nuclear magnetic moment	Resonances, nuclear induction	Nuclear magnetic resonance effects, nuclear induction
Nuclear absorption of gamma rays	Neutron emission	Nuclear photoeffect
Electromagnetic fields acting on nuclear spin	Hyperfine structure	Nuclear spin effect
Additional high-frequency field in nuclear magnetic resonance	Increased upper level population, increase in nuclear signal	Overhauser effect
Recoil in neutron capture and particle emission	Change in the state of chemical bonding of radioactive atoms	Szilard-Chalmers effect
Overheating in reactor coolant	Increase in radioactivity	Void effect
Fast neutrons acting on graphite	Defects in graphite lattice	Wigner effect

Lit.: [6d, 6e, 17, 22d, 41, 69, 70 Vol. II, 90, 96, 100, 101, 114]

Table 39. Optics (O)

Cause	Effect	Name
Diffraction of light (electromagnetic radiation) at complementary screens	Complementary diffraction patterns	Babinet's principle
Standing ultrasonic waves in liquids	Density modulation acts as diffraction grating	Debye-Sears effects
Light propagation in a moving optical medium	Light transported by the medium, shift in interference fringes	Fizeau effect
Total reflection of light, e.g., in a prism	Beam shift	Goos-Hähnchen effect
Double reflection of light from glass plates at the Brewster angle	Linearly polarized light	Malus's experiment
Velocity gradients in a flowing liquid	Birefringence due to orientation of anisotropic particles	Maxwell effect
Fast movement of a light source	Frequency shift	Optical Doppler effect, Stark-Doppler effect
Absorption of frequency-modulated light, e.g., in gases	Gas warming and cooling, modulated heat source	Optoacoustic effect, photoacoustic effect
Intense light acting on a transparent dielectric	Refractive index change	Photorefractive effects
Rotation of an interferometer, e.g., a ring interferometer	Shift in interference fringes dependent on state of rotation	Sagnac effect

Lit.: [6c, 9, 10, 16, 22c, 38 Vol. II, Vol. III, 40, 51, 61 Vol. III, 63, 65, 96, 98, 100, 101, 106, 114, 119]

Table 40. Photoelectricity (P)

Cause	Effect	Name
Light acting on one of two electrodes in an electrolyte	Potential difference between the electrodes	E. Becquerel effect
Light acting on a gas in an electric field	Incomplete ionization	Branley-Lenard effect
Light acting on a negatively charged metal plate	Plate discharged, electron release	Hallwachs effect, optoelectric effect
Light acting on a spark discharge	Reduction in breakdown voltage	Hertz effect

Table 40 (continued)

Cause	Effect	Name
Light acting on a metal surface	Electron release	Photoelectric effect, external photoelectric effect, internal photoelectric effect, opto-electric effect, Maggi effect, normal photoelectric effect
Light acting on a thin metal film	Resonant-type dependence of electron emission on light energy	Selective photoelectric effect

Lit.: [2, 3, 6c, 7, 27, 50 Vol. II, 2.1, Vol. 4, part 3, 53, 94, 100, 104, 111, 114,]

Table 41. Plasma (Pl)

Cause	Effect	Name
Flow of a conducting liquid or gas in a transverse magnetic field	Hall effect, transverse potential difference	Magnetohydrodynamic effect
Light incident on a discharge tube	Change in voltage and/or discharge current	Optogalvanic effect, optovoltaic effect
Magneti field applied to a current-carrying plasma	Plasma contraction	Pinch effect

Lit.: [6d, 22d, 90, 100, 114]

Table 42. Photoelectric effects in semiconductors (PS)

Cause	Effect	Name
Illumination of a p-n junction	Shift in diode characteristics	Barrier-layer photoelectric effect, p-n photoelectric effect, photovoltaic effect
Light falling on a plate thinner than the wavelength	Polaritons, transverse or longitudinal bulk phonons	Berremann effect
Illumination of cuprite crystals	Photo-EMF dependent on light direction	Crystal photoelectric effect
Light absorption in the surface layer of a semiconductor	Diffusion current	Dember effect

(continued on page 116)

Table 42 (continued)

Cause	Effect	Name
Illumination of a semiconductor specimen under voltage	Conductivity change	Internal photoelectric effect, photoconduction
Illumination of a semiconductor acting as the dielectric in a capacitor	Capacitance change	Photocapacitative effect
Illumination of a semiconductor at low temperatures	Change in dielectric parameters	Photodielectric effect
Magnetic field perpendicular to incident light	Change in Hall voltage	Photo-Hall effect, PEM (photoelectromagnetic) effect, photogalvanomagnetic effect

Lit.: [2, 3, 6d, 8, 17, 22d, 27, 28, 49, 53, 63, 65, 78, 94, 96, 100, 101, 104, 110, 114]

Table 43. Relativistic physics (R)

Cause	Effect	Name
Gravitation	Perihelion precession, light deflection, gravitational frequency shift	Einstein effects (Pound-Rebka experiment, Shapiro experiments, time effects)
Gravitational force acting on a balance bearing different materials	No tilt or twisting, i.e., quality of inertial and gravitational masses	Eötvös experiment
Propagation of light in a moving system	Change in interference on interferometer rotation	Michelson's experiment
Rotating gyroscope in a rotating hollow body	Gyroscope precession	Thirring-Lense effect
Motion of a dipole	Dipole should exert torque on itself	Trouton-Noble experiment

Lit.: [6c, 10, 14, 38 Vol. II, 40, 50 Vol. II, part 1, Vol. 5, part 2, 59, 111, 118a, 120]

Table 44. Superconductivity (S)

Cause	Effect	Name
Anisotropic crystal	Direction-dependent Cooper pair production	Anisotropy effect
Unstable domains in superconductor	Dependence of current-carrying capacity on instabilities	Degradation effect
Superconductors separated by a thin insulating layer	Tunnelling of single electrons, tunnelling current	Electron-tunnelling effect
Different isotopes	Changes in the transition temperature and critical magnetic field	Isotope effect
Compensation of the polarisation of the conduction electrons by magnetic field in a ferromagnetic material	Induced superconduction	Jaccarino-Peter effect
Potential difference across a tunnel junction	High-frequency alternating current dependent on voltage across junction	AC Josephson effect
No potential difference across a tunnel junction	Flow of a superconducting Cooper-pair current	DC Josephson effect
External magnetic field acting on a superconductor	Inner part of the superconductor remains field-free	Meissner-Ochsenfeld effect, Meissner effect, penetration effect
Material inhomogeneities	Existence of superconducting domains above the transition temperature	Nucleation effect
Exceeding the critical temperature	No loss of superconductivity	Overheating effect
Trapping centers in superconductor	Maximum in the critical current density as a function of magnetic field	Peak effect
Continity of order parameters in a normal-superconductor boundary layer	Existence of Cooper pairs in a normally conducting layer adjoining a superconductor	Proximity effect (coupling effect)
External magnetic field	Magnetic field screening by eddy currents	Screening effect, Maxwell-Lippmann effect
Falling below the critical temperature	No onset of superconductivity	Undercooling effect
Incorporation of defect atoms into a semiconductor	Change in superconductor transition temperature	Valency effect

Lit.: [3, 6d, 11, 15, 20, 21, 22d, 27, 30, 41, 44, 50 Vol. 4, part 1, 100, 114]

Table 45. Scattering, elastic (Sc)

Cause	Effect	Name
Light scattering at particles comparable with or larger than the wavelength	Characteristic scattering patterns, definite states of polarization	Mie scattering
Light scattering at nuclei	Nuclear level excitation	Nuclear resonance scattering
Light scattering by particles which are small relative to the wavelength, e.g., in gases	Transmitted light redder scattered light bluer	Rayleigh scattering
Light scattering at atoms or molecules	Shell electron excitation	Resonance scattering
Light scattering at elastically bound electrons	Electrons act as dipoles radiating with the same frequency	Thomson scattering
Light scattering at submicroscopic particles in solution	Tyndall cone, with the transmitted light redder and the scattered light bluer	Tyndall effect

Lit.: [6d, 7, 22d, 38 Vol. II, III, IV, VIII, 41, 46, 98, 100, 109, 114]

Table 46. Scattering, inelastic (Sc)

Cause	Effect	Name
Photon scattering at acoustic oscillations (phonons) in solids and liquids	Characteristic lines and doublets	Brillouin scattering (Brillouin effect)
Photon scattering at electrons	Longer-wavelength photons, recoil electrons	Compton effect
Photon scattering at nuclei	Electron-positron pairs	Pair production
Photon scattering at vibrating atoms, molecules, or atomic or molecular lattices	Lines characteristic of the molecule	Raman effect

Lit.: [6d, 7, 22d, 38 Vol. III, IV, VIII, 41, 42, 45, 46, 98, 100, 109, 114]

Table 47. Scattering, particle (Sc)

Cause	Effect	Name
Scattering of high-energy positrons at electrons		Bhabha scattering
Scattering of γ rays at the nuclear potential		Delbrück scattering
Polarized-electron scattering at a Coulomb potential		Mott scattering
Scattering of light-particles (H, He) at potentials containig attractive and repulsive parts	Rainbow in analogy to scattering of light at liquid droplets	Rainbow effect
Charged-particle scattering at charged particles		Rutherford scattering (Coulomb scattering)

Lit.: [6d, 6e, 7, 16, 17, 22d, 98, 100, 109, 114]

Table 48. Semiconductors (SC)

Cause	Effect	Name
Occupation of lower edge of conduction band by donors	Shift in absorption edge	Burstein effect
Temperature gradient in semiconductor	Nonequilibrium phonon distribution	Drag effect (transport effect)
Electric field in semiconductor	Local change in charge distribution	Field effect
External electric field acting on semiconductor	Shift in optical absorption edge to higher energies	Franz-Keldysh effect
Electric and magnetic fields in semiconductors	Temperature and voltage differences, resistance change	Galvanomagnetic effects
Lack of long-range order in a semiconductor	Broadening of the conduction and valence bands	Hopping effect
Potential difference between semiconductor and metal specimen	Trapping	Johnsen-Rahbeck effect
Semiconductor in magnetic field	e.g., Change in the excitation of electromagnetic waves	Magnetoplasma effect
Mechanical stress on a p-n junction	Change in current-voltage characteristics	Piezojunction effect

(continued on page 120)

Table 48 (continued)

Cause	Effect	Name
Mechanical stresses in solids	Change in electrical resistance	Piezoresistance effect
Electric field acting on a semiconductor	Increase in electrical conductivity	Poole-Frenkel effect
Illumination of silver halide crystals	Depostion of colloidal silver	Printout effect
High field strengths at a p-n junction	Internal field emission	Schottky effect
Electric and magnetic fields acting on a semiconductor	Increased charge-carrier recombination near crystal surface	Suhl effect

Lit.: [3, 8, 27, 28, 43, 49, 53, 65, 94, 100, 104, 110, 114]

Table 49. Semiconductor components (SCC)

Cause	Effect	Name
High voltage difference across the space-charge zone in a p-n junction	New charge carriers produced by collisional ionization	Avalanche effect
Several energy minima in semiconductor conduction band	Negative differential conductivity, high-frequency oscillations	Gunn effect
Temperature coefficient of conductivity for a semiconductor positive or negative	Nonlinear current-voltage characteristic	Hot conductor effect or cold conductor effect
Stress in an amorphous semiconductor	State persists when external stress vanishes	Memory effect
Stress in an amorphous semiconductor	Sudden change in conductivity	Ovshinsky effects, switching effect
Current flow through a metal-semiconductor contact	Current flow only in one direction	Rectifier effect (barrier or barrier-layer effect)
Light acting on amorphous silicon	Change in dark conductivity and photoconductivity	Staebler-Wronski effect
Directional charge carrier injection into semiconductor	Increase in injection current	Transistor effect (this is related to the following: early effect, emitter-dip effect, Kirk effect)
Microscopically small oppositely facing p-n junctions in sintered silicon carbide	Nonlinear voltage-dependent resistance	Varistor effect
Increased leakage current on p-n junction	Electron tunneling from valency band into conduction band, steep current increase	Zener effect

Lit.: [6d, 22d, 28, 53, 63, 64, 94, 100, 104, 110, 114]

Table 50. Solidity (So)

Cause	Effect	Name
Various	Hysteresis in the case of alternating stresses	Aftereffects (relaxation effects)
Plastic deformation	Change in E modulus	Bauschinger effect
Metal rod bending	Change in hydrogen concentration, inelastic extension	Gorsky effect
Mechanical deformation and solvent	Increase in plasticity of ionic crystals	Ioffe effect
Foreign atoms in a material	Change of the hardening curve and the elastic limit	Portevin-Le Chatelier effects
Torsion in a long rod	Rod lengthening	Poynting effect
Molecular surface films on a material	Reduction in hardness and ductility	Rehbinder effect
Carbon in iron	Occupation of special lattice sites	Snoek effect
Stretching high-polymer fibers	Hardness increased, necking in fiber	Telescope effect

Lit.: [37, 52a, 100, 114]

Table 51. Solid state (SS)

Cause	Effect	Name
Sound absorption in semiconductor	Sound wave amplification	Acoustoelectric effect
Magnetic field and microwave field acting on a single metallic crystal at low temperatures	Cyclotron resonance (see Dictionary of Effects)	Azbel-Kaner effect, cyclotron resonance
Emergence of a diffracted X-ray beam	Anomalously low absorption	Borrmann effect
Motion of protons between planes occupied by ions	Preferential proton motion in channels	Channelling effect
Impurity clouds in a crystal	Blocking of dislocation planes	Cottrell effect
Particles (neutrons) acting on a crystal lattice	Particle interaction cross section dependent on crystal structure	Crystal effect

(continued on page 122)

Table 51 (continued)

Cause	Effect	Name
Strong magnetic field acting on a single crystal at low temperatures	Electron oscillations at the Fermi surface	De Haas-van Alphen effect
Effect of lattice on the electron states in a defect in the crystal lattice	Lifting of degeneracy for the defect site states	Jahn-Teller effect
Diffraction of diffuse electrons in a thick crystal layer	Production of characteristic lines and cones	Kikuchi effect
Alloy components with different diffusion constants	Alloy component diffusion, shift in separating surfaces	Kirkendall effect
Change in phonon wave-number vector in a material	Change in phonon-dispersion curve	Kohn effect
Incorporation of magnetic foreign atoms into metals	Resistance minimum at low temperatures	Kondo effect
X-ray diffraction at an anti-cathode crystal	Inherent emission, which occurs in double cones relative to the lattice planes	Kossel effect
X-rays acting on crystal wafers	Characteristic X-ray interferences	Laue effect (Laue method)
Ultrasound absorption in a material at low temperatures in a homogeneous magnetic field	Absorption oscillation	Magnetoacoustic effect
Martensite transformations in alloys at various temperatures	Mechanical state memory	Memory effect
X-ray diffraction at a crystal	Diffracted beam acting as primary beam	Renninger effect
Drifting of ions from lattice sites	Deposition on surface	Schottky effect
Magnetic field acting on a single crystal at low temperatures	Oscillation in electrical resistance	Shubnikov-de Haas effect
Absorption and reflection of high-frequency radiation (MHz) as a function of temperature and magnetic field	Oscillation in absorption, reflection, and surface impedance	Size effects (Gantmacher or radio frequency effect)
Magnetic field and sample surface mutually at an angle, microwave field perpendicular to sample	Spiral paths, change in static conductivity	Sondheimer effect

Lit.: [1, 3, 6d, 21, 25, 27, 50 Vol. 4, 71, 75, 94, 100, 102, 104, 110, 111, 114]

Table 52. Synergetic effects (Sy)

Cause	Effect	Name
Nonlinear heat transfer	Hexagonal cells and polygonal rolls	Bénard effect
Change of velocity of fluid between layers	Instabilities at the boundary layer	Kelvin-Helmholtz effect
Placing of a layer of heavy liquid above a lighter one	Disturbation of the labil equilibrium	Rayleigh-Taylor effect
Couette flow between rotating cylinders	Generation of a regular pattern at a critical angular velocity	Taylor effect

Lit.: [102a, 103a and b, 122a, 116a]

Table 53. Thermoelectricity (Te)

Cause	Effect	Name
Electrical current or heat flux through homogeneous conductor	Temperature difference, potential difference	Benedicks effects, Thomson effects
Current flow through an anisotropic crystal	Heat or cold	Internal Peltier effect, Bridgman effect
Electric current or heat flux through two different connected conductors	Heat or cold at the points of junction, potential difference between the open ends of the conductors	Peltier effect Seebeck effect
Shift in phonon Fermi surface	Phonon flux	Phonon drag effect, Gurevich effect

Lit.: 1, 3, 6b, 18, 22d, 23, 30, 38 Vol. VIII, 43, 50 Vol. 4, part 4, 53, 62, 71, 86 Vol. XVI, part 1, 100, 114, 117]

Table 54. Thermodynamics and kinetics (TK)

Cause	Effect	Name
Gravitational force on a binary mixture containing substances with very different boiling points	Heavy vapor phase descends below the liquid surface	Barotropic phenomenon
Osmotic pressure acting on an electrical potential, etc.	Various	Colligative effects, osmotic effects
Concentration difference in a mixture	Heat flux	Diffusion thermo-effect
Diffusion of two substances	Temperature gradient	Dufour effect
Increase of temperature	Increase of volume of a gas	Gay-Lussac experiment
Isotope mixture	Differences in diffusion and reaction rate	Isotope effect
Adiabatic expansion of real gases	Temperature reduction	Joule-Thomson effect (throttle effect)
Flow of gas through a capillary linking two vessels with differing temperatures	Flow in opposite direction to temperature gradient	Knudsen effect, thermal effusion
Temperature gradient in a mixture	Concentration gradient	Ludwig-Soret effect
Surfaces differing in temperature in vacuum	Pressure difference between the surfaces	Radiometer effect
Addition of electrolytes to a solution	Change in non-electrolyte solubility	Salting-out or salting-in effect
Magnetic field applied to a gas	Change in thermal conductivity	Senftleben effect
Mean free path of the same order as the vessel size and decreasing pressure	Decrease in thermal conductivity in a vessel filled with a fine powder	Smoluchowski effect

Lit.: [6a, 20, 22a, 23, 35, 35, 50 Vol. III, part 1 and 2, 96, 100, 101, 102, 113, 114]

Table 55. Thermomagnetism (Tm)

Cause	Effect	Name
magnetic field	Transverse potential difference	First Ettingshausen-Nernst effect
Heat flux passing through a conductor in a transverse	Transverse temperature difference	First Righi-Leduc effect

Table 55 (continued)

Cause	Effect	Name
Absorption of heat radiation in a sample in a transverse magnetic field	Transverse potential difference due to longitudinal temperature difference	OEN effect: optically induced Ettingshausen-Nernst effect
Heat flux passing through a conductor in a longitudinal magnetic field	Longitudinal potential difference	Second Ettingshausen-Nernst effect
Heat flux passing through a conductor in a transverse magnetic field	Longitudinal temperature difference	Second Righi-Leduc effect, Maggi-Righi-Leduc effect
Heat flux passing through a conductor in a longitudinal magnetic field	Change in thermal resistance	(No special name)
Heat flux passing through a conductor in a transverse magnetic field	Thermo-EMF between magnetized and unmagnetized conductors	(No special name)

Lit.: [1, 3, 6b, 20, 22d, 23, 30, 38, 43, 49, 50 Vol. 4, part 4, 53, 62, 71, 86 Vol. XIII, 89, 94, 100, 104, 110, 114]

Chronology

Effects and major events in physics (selection)

1756 Leidenfrost phenomenon
1777 Lichtenberg figures
1801 Volta effect
1802 First Gay-Lussac experiment
1807 Electrophoresis, Second Gay-Lussac experiment
1808 Malus's experiment
1818 Light diffraction (Fresnel)
1820 Deflection of a magnetic needle by an electric current (Oerstedt), Ampère's laws
1822 Seebeck effect
Diffraction at a grating (Fraunhofer)
1824 Arago's experiment
1825 Radiometer effect
1826 Relation between current, resistance, and voltage: Ohm's law
1827 Brownian molecular movement
1831 Electromagnetic induction (Faraday)
1832 Coriolis force
1833 Lenz's rule
1834 Peltier effect
1839 E. Becquerel effect
1841 Joule effect (current heating)
1842 Doppler effect
Joule effect (magnetostriction)
1845 Faraday effect
1847 Matteucci effect
Energy production (Helmholtz)
1849 Experimental estimation of the velocity of light (Fizeau)
1850 Foucault's experiment
Fizeau's experiment
1852 Wertheim effect
Joule-Thomson effect
1856 Thomson effects
Ludwig-Soret effect
1858 Wiedemann effect
1859 Flow current (Quincke)
1860 Maxwellian velocity distribution
1861/
1862 Maxwell's electrodynamic equations
1865 Villary effect
Second law of thermodynamics (Clausius)
1866/
1867 Dynamoelectric principle (W. von Siemens)
1869 Tyndall effect
1872 Dufour effect
1873 Photoconduction
1875 Kerr effect
Rectifier effect
1876 Electropapillarity (Lippmann)
Magnetooptic Kerr effect
Barrier-layer photoelectric effect

1878 Dorn effect
Rowland effect
1879 Hall effect
Edison effect
Ludwig-Soret effect
1880 Piezoelectric effect
1881 Photoacoustic effect
Michelson's experiment
1884 Kundt effect
1885 First series law for spectral lines (J. J. Balmer)
Edison effect
1886 Ettingshausen effect
Ettingshausen-Nernst effect
Hertz effect, Bauschinger effect
1887 Nernst effect, Righi-Leduc effect
Lenard-Hertz effect, Hertz effect
Michelson-Morley experiment
1888 Hallwachs effect, Hertz effect
Röntgen-Eichenwald experiment
1890 Lenard effect, Hopkinson effect
1892 Lenard effect
1893 Pockels effect
1894 Selective photoelectric effect
1895 Discovery of X rays (W. C. Röntgen)
1896 Zeeman effect
Discovery of radioactivity (H. Becquerel)
1898 Voigt effect, Macaluso-Corbino effect
1900 Rayleigh scattering
Quantum hypothesis (M. Planck)
1901 Observation of light pressure (Lebedev)
1902 Majorana effect
1903 Thomson scattering, skin effect
Trouton-Noble experiment
Richardson effect
1905 Stark-Doppler effect (optical Doppler effect)
Special relativity theory (A. Einstein)
Photon model of light (A. Einstein)
1906 Barotropic phenomenon
J. Becquerel effect
1907 Cotton-Mouton effect
1908 Mie effect
α radiation recognized as helium nuclei
(E. Rutherford)
Cooling effect
1910 Knudsen effect
1911 Beck effect, Corbino effect
Superconduction effect
Smoluchowski effect
Light deflection
Rutherford's atomic model
1912 Debye effects, Laue effect
Paschen-Back effect

1913 Onnes effects (Kammerlingh-Onnes)
Stark effect
Frank-Hertz experiment, Sagnac effect
N. Bohr's atomic model
Measurement of the elementary electric charge
(R. A. Millikan)
1915 Barnett effect
Einstein-de Haas effect
General relativity theory (A. Einstein)
1916 Benedicks effects, dissymmetry effect
1917 Barkhausen effect
1918 Thirring-Lense effect
Magnetocaloric effect
Schottky effect
1919 Shielding effect, Barkhausen effect
Eötvös effect, Pohl effect
First artificial nuclear transmutation
(Rutherford)
Deflection of light in the Sun's gravitational field (A. S. Eddington)
1920 Ramsauer effect, Gudden-Pohl effect
1921 Stern-Gerlach effect (suggestion)
1922 Brillouin effect
Stern-Gerlach effect
1923 Prediction of the Raman effect by Smekal
Langmuir effect
Compton effect
Debye-Hückel theory of electrolyte conduction (braking effects)
Portevin-le Chatelier effect
Photoelectric effect, internal
Lossew effect
Wave property of particles (L. de Broglie)
1924 Hanle effect
Ioffe effect
1925 Development of quantum mechanics (W. Heisenberg)
Exclusion principle (W. Pauli)
Bridgman effect
1926 Auger effect
Field effect (field electron emission)
Franck-Condon principle
Wave mechanics (E. Schrödinger)
1927 Wien effect
Back-Goudsmith effect
Electron-beam interferences (C. J. Davisson, L. H. Germer, G. P. Thomson, and A. Reid)
Uncertainty relation (W. Heisenberg)
1928 Tunnelling effect (G. Gamow)
Debye-Falkenhagen effect
Kikuchi effect
Penning effect
Raman effect
1929 Hubble effect
1930 Weissenberg effect

1931 Dissociation-voltage effect
Dember effect
de Haas-van Alphen effect
1932 Johnson-Rahbeck effect
Discovery of the positron (C. D. Anderson)
Discovery of the neutron (J. Chadwick)
Construction of the atomic nucleus from neutrons and protons (W. Heisenberg and D. Ivanenko)
1933 Meissner-Ochsenfeld effect
Delbrück scattering
1934 Nuclear photoeffect
Kossel effect
Szilard-Chalmers effect
Cherenkov effect
Photo-Hall effect
Theory of β decay (E. Fermi)
Discovery of artificial radioactivity (I. and F. Joliot-Curie)
1935 Prediction of mesons (H. Yukawa, first published in Japan in 1934)
Hughes effect
Mass formula (Weizsäcker, Bethe)
Déchêne effect
Ueling effect
Schottky effect
1936 Destriau effect
1937 Jahn-Teller effect
Renninger effect
1938 Snoek effect
Nuclear fission (O. Hahn, F. Strassmann)
1939 Hot conductors
Kirlian effect
1940 Anomalous skin effect
1942 Controlled chain reaction (E. Fermi)
Kirkendall effect
1943 Nucleation effect
1945 Explosion of two atom bombs over Hiroshima and Nagasaki
1946 Development of transistor (W. B. Shockley)
1947 Holography (D. Gabor)
Lamb shift
1948 Transistor theory (J. Bardeen, W. H. Brattain)
1949 Suhl effect
1950 Sondheimer effect
Isotope effect (superconductivity)
Schoch effect
1951 Foldy effect
Haas effect
1952 Shpol'skii effect
1953 Overhauser effect
Avalanche effect
Early effect
Smith-Purcell-Salisbury effect
1954 Burstein effect
1955 Observation of antiprotons (O. Chamberlain, E. Segré)

1956 Theory of parity violation in weak interaction (C. N. Yang, T. D. Lee)
Azbel-Kaner effect
1957 Observation of parity violation (C. S. Wu et al.)
1958 Mössbauer effect
Theory of masers and lasers (C. H. Townes, N. Basow, A. Prochorov)
Size effects
Franz-Keldysh effect
Formulation of the VA theory to the weak interaction (E. C. G. Sudarshan, R. Marshak, M. Gell-Mann, R. Feynman)
1960 Lasers (T. H. Maiman)
Electron tunnelling effect
Pound-Rebka experiment
1961 Symmetry of baryons and mesons (M. Gell-Mann, Y. Ne'eman)
Slope effect
1962 Josephson effects
Gantmacher effect
Kirk effect
Self-focussing
1963 Berremann effect
Gunn effect
1964 Kondo effect
Experimental observation of CP violation
1966 Ovshinsky effects
1967 Magnetic memory effect (Bobeck)
1968 Emitter-dip-effect
Ovshinsky effects
Diffraction filter-ray bundle effect
1969 Fano effect
1971 Schadt-Helfrich effect
Hafele-Keating experiment
1972 Memory effect
Rehbinder effect
1973 Discovery of neutral weak currents (Faissner)
1974 Discovery of narrow resonances (J or Ψ particles, Ting and Richter)
1975 Wiegand effect
1976 Memory effect
Switching effect
1980 v. Klitzing effect (quantum Hall effect)
1983 Fractional quantum Hall effect
1984 Jaccarino-Peter effect

Literature

Textbooks/Monographs

[1] ABRIKOSSOW, A. A.: *Einführung in die Theorie normaler Metalle* (J. MERTSCHING, ed.). Vieweg, Brunswick 1977.

[2] AUTH, J., GENZOW, D., and HERMANN, K. H.: *Photoelektrische Erscheinungen*. Vieweg, Brunswick 1977.

[3] BECKER, R., and SAUTER, F.: *Theorie der Elektrizität*, Vol. 3. Teubner, Stuttgart 1969.

[4] BECKER, R., and DÖRING, W.: *Ferromagnetismus*. Springer, Berlin 1939.

[5] BERBER, J., KACHER, H., and LANGER, G.: *Physik in Formeln u. Tabellen*. Verlag Handwerk und Technik, Hamburg 1975.

[6] BERGMANN, L., and SCHAEFER, C.: *Lehrbuch der Experimentalphysik*. Walter de Gruyter, Berlin.

[6a] BERGMANN/SCHAEFER: Vol. 1, 9th ed. 1974.

[6b] BERGMANN/SCHAEFER: Vol. 2, 6th ed. 1974.

[6c] BERGMANN/SCHAEFER: Vol. 3, 7th ed. 1978.

[6d] BERGMANN/SCHAEFER: Vol. 4, 1975.

[7] BETHGE, K.: *Quantenphysik*. Bibliographisches Institut, Mannheim 1978.

[8] BERGH, A., and DEAN, P. J.: *Lumineszenzdioden*. Hüthig, Heidelberg 1976.

[9] BORN, M.: *Optik*. 3rd ed. Springer, Berlin-Heidelberg-New York 1972.

[10] BORN, M.: *Die Relativitätstheorie Einsteins*. 5. ed. Springer, Berlin-Heidelberg-New York 1969.

[11] BUCKEL, W.: *Supraleitung*. 3rd ed. Physik-Verlag, Weinheim 1984.

[12] BUTLER, J. A. V.: *Electrical Phenomena at Interfaces*. Methuen, London 1951.

[13] BURCHAM, W. E.: *Elements of Nuclear Physics*. Longman, Harlow 1979.

[14] DAUTCOURT, G.: *Relativistische Astrophysik*. Vieweg, Brunswick 1972.

[15] ENGELHAGE, D., and DALLWITZ, L.: *Grundlagen supraleitender elektronischer Bauelemente und Schaltungen*. Akadem. Verlagsgesellschaft, Leipzig 1978.

[16] FLEISCHMANN, R.: *Einführung in die Physik*. 2rd ed. Physik-Verlag, Weinheim, 1980.

[17] FINKELNBURG, W.: *Einführung in die Atomphysik*. 11th/12th ed. Springer, Berlin-Heidelberg-New York 1976.

[18] FELDKELLER, E.: *Dielektrische und magnetische Materialeigenschaften*. Vol. I. and II. Bibliographisches Institut, Mannheim 1974.

[19] FERETTI, M.: *Laser, Maser, Hologramme*. Franzis, Munich 1977.

[20] FRITSCH, G.: *Transport*. Akadem. Verlagsgesellschaft, Leipzig 1979.

[21] GENZEL, L.: *Die feste Materie*. Umschau Verlag, Frankfurt 1973.

[22] GRIMSEHL, B. G.: *Lehrbuch der Physik*. Teubner, Leipzig.

[22a] GRIMSEHL, Vol. 1, 1977.

[22b] GRIMSEHL, Vol. 2, 1980.

[22c] GRIMSEHL, Vol. 3, 1978.

[22d] GRIMSEHL, Vol. 4, 1975.

[23] DE GROOT, S. R.: *Thermodynamik irreversibler Prozesse*. Bibliographisches Institut, Mannheim 1960.

[24] HAMANN, C. R., and VIELSTICH, W.: *Elektrochemie I. and II*. Verlag Chemie, Weinheim 1975 and 1981.

[25] HAUSSÜHL, S.: *Kristallgeometrie*. Verlag Chemie, Weinheim 1979.

[26] HECK, C.: *Magnetische Werkstoffe und ihre technische Anwendung*. Hüthig, Heidelberg 1975.

[27] HELLWEGE, K.-H.: *Einführung in die Festkörperphysik*. Springer, Berlin-Heidelberg-New York 1976.

[28] HESSE, K.: *Halbleiter I*. Bibliographisches Institut, Mannheim 1974.

[29] HUANG, D.: *Statistische Mechanik*. Vol. I, II and III. Bibliographisches Institut, Mannheim 1964.

[30] JUSTI, E.: *Leitfähigkeit und Leitungsmechanismus fester Stoffe*. Vandenhoek & Ruprecht, Göttingen 1948.

[31] KAMPCZYK, W., and RÖSS, E.: *Ferritkerne*. Siemens AG, Berlin 1978.

[32] KARPMANN, V. I.: *Nichtlineare Wellen in dispersiven Medien*. Vieweg, Brunswick 1977.

[33] KIEMLE, H., and RÖSS, E.: *Einführung in die Technik der Holographie*. Akadem. Verlagsgesellschaft, Leipzig 1969.

[34] KNELLER, E.: *Ferromagnetismus*. Springer, Berlin-Heidelberg-New York 1962.

[35] KORTÜM, G., and LACHMANN, H.: *Einführung in die chemische Thermodynamik*. 6th ed. Verlag Chemie, Weinheim 1972, and Vandenhoek & Ruprecht, Göttingen 1981.

[36] KRESTEL, E. (ed.): *Bildgebende Systeme für die medizinische Diagnostik*. Siemens AG, Berlin 1980.

[37] LAMBECK, M.: *Barkhausen-Effekt und Nachwirkung in Ferromagnetika*. Walter de Gruyter, Berlin 1971.

[38] LANDAU, L., and LIFSCHITZ, E.: *Lehrbuch der Theoretischen Physik*. Akademie-Verlag, Berlin.

[38a] LANDAU/LIFSCHITZ, Vol. 1, 1979.

[38b] LANDAU/LIFSCHITZ, Vol. 2, 1977.

[38c] LANDAU/LIFSCHITZ, Vol. 3, 1978.

[38d] LANDAU/LIFSCHITZ, Vol. 4, 1979.

[38e] LANDAU/LIFSCHITZ, Vol. 5, 1978.

[38f] LANDAU/LIFSCHITZ, Vol. 6, 1978.

[38g] LANDAU/LIFSCHITZ, Vol. 7, 1977.

[38h] LANDAU/LIFSCHITZ, Vol. 8, 1974.

[38i] LANDAU/LIFSCHITZ, Vol. 9, 1980.

[38j] LANDAU/LIFSCHITZ, Vol. 10, 1983.

[39] VON LAUE, M.: *Geschichte der Physik*. Ullstein, Berlin 1966.

[40] VON LAUE, M.: *Die Relativitätstheorie*. Vol. I and II. Vieweg, Brunswick 1965.

[41] LEIGHTON, R. B.: *Principles of Modern Physics*. McGraw Hill, London 1959.

[42] LETOCHOW, W. S.: *Laserspektroskopie*. Vieweg, Brunswick 1977.

[43] LINDNER, H.: *Grundriß der Festkörperphysik*. Vieweg, Brunswick 1979.

[44] LYNTON, E. A.: *Supraleitung*. Bibliographisches Institut, Mannheim 1966.

[45] MATOSSI, F.: *Der Raman-Effekt*. Vieweg, Brunswick 1959.

[46] MAYER-KUCKUK, T.: *Atomphysik*. Teubner, Stuttgart 1977.

[47] MAYER-KUCKUK, T.: *Kernphysik*. Teubner, Stuttgart 1979.

[48] KLEEN, W., and MÜLLER, R.: *Laser*. Springer, Berlin-Heidelberg-New York 1969.

[49] MIERDEL, G.: *Elektrophysik*. Hüthig, Heidelberg 1977.

[50] *Müller-Pouiletts Lehrbuch der Physik*. Vol. 1 to Vol. 5 (14 parts), 11th ed. Vieweg, Brunswick 1924 – 1934, Vol. I to III, 9th ed. 1888 – 1902.

[51] NEUFELDT, S.: *Chronologie Chemie 1800 – 1970*. Verlag Chemie, Weinheim 1977.

[52] ORTON, J. W.: *Gunn-Effekt-Halbleiter*. Hüthig, Heidelberg 1973.

[52a] PAUFLER, P., and SCHULZE, G. E.: *Physikalische Grundlagen mechanischer Festkörpereigenschaften*. Vieweg, Brunswick 1978.

[53] PAUL, R.: *Halbleiterphysik*. Hüthig, Heidelberg 1975.

[54] POGGENDORF, D.: *Bibliographisches Wörterbuch zur Geschichte der exakten Wissenschaften*. Vol. I (Leipzig 1863) to Vol. VII b. Akademie-Verlag, Berlin.

[55] RAMSAUER, C.: *Grundversuche der Physik in historischer Darstellung*. Springer, Berlin 1953.

[56] RÖSS, D.: *Laser und Lichtoszillatoren*. Akadem. Verlagsgesellschaft, Leipzig 1966.

[57] ROSENBERGER, D.: *Technische Anwendung des Lasers*. Springer, Berlin-Heidelberg-New York 1975.

[58] SACHSE, H.: *Ferroelektrika*. Springer, Berlin 1956.

[59] SEXL, R. U., and URBANTKE, H. K.: *Gravitation und Kosmologie*. Bibliographisches Institut, Mannheim 1975.

[60] SOMMERFELD, A.: *Atombau und Spektrallinien*. Vol. I and II. Vieweg, Brunswick 1978.
[61] SOMMERFELD, A.: *Vorlesungen über Theoretische Physik*. Vol. I to VI. Akadem. Verlagsgesellschaft, Leipzig 1978.
[62] SCHRÖDER, H., and SCHULTE, C.: *Elektrische Leitungsvorgänge* (F. BERGMANN, H. SCHRÖDER, ed.). Diesterweg/Salle/Sauerländer, Frankfurt 1979.
[63] SPENKE, E.: *Elektronische Halbleiter*. Springer, Berlin-Heidelberg-New York 1965.
[64] STAHL, K. J., and MIOSGA, G.: *Infrarottechnik*. Hüthig, Heidelberg 1980.
[65] TEICHMANN, H.: *Halbleiter*. Bibliographisches Institut, Mannheim 1961.
[66] TEICHMANN, H.: *Einführung in die Atomphysik*. 3rd ed. Bibliographisches Institut, Mannheim 1966.
[67] WEBER, H., and HERZIGER, G.: *Laser. Grundlagen und Anwendungen*. Physik-Verlag, Weinheim 1972.
[68] URLAUB, J.: *Röntgen-Analyse*. Siemens AG, Berlin 1974.
[69] WESTPHAL, W. H.: *Physik*. Springer, Berlin-Heidelberg-New York 1970.
[70] WEITZEL, W.: *Lehrbuch der Theoretischen Pyhsik I und II*. Springer, Berlin 1955/1958.
[71] ZIMAN, J. M.: *Prinzipien der Festkörpertheorie*. Harri Deutsch, Frankfurt 1975.

Dictionaries/Handbooks

[72] *Lexikon der Physik*. Vol. 1 to 10. Deutscher Taschenbuch Verlag, Munich 1970.
[73] *Lexikon der Physik*. (H. FRANKE, ed.). Francksche Verlagsbuchhandlung, Stuttgart 1950.
[74] *Meyers Lexikon der Physik*. Bibliographisches Institut, Mannheim 1973.
[75] *Fachlexikon abc Physik*. Harri Deutsch, Frankfurt 1974.
[76] *Physikalisches Wörterbuch*. Springer, Berlin 1952.
[77] *Physikalisches Handwörterbuch*. Springer, Berlin 1924.
[78] *Physikalisches Taschenbuch*. Vieweg, Brunswick 1976.
[79] *Atlas zur Atomphysik*. Deutscher Taschenbuch Verlag, Munich 1976.
[80] *Atlas zur Astronomie*. Deutscher Taschenbuch Verlag, Munich 1973.
[81] *Herder Lexikon der Physik*. Herder, Freiburg 1979.
[82] *Herder Lexikon für Naturwissenschaftler*. Herder, Freiburg 1979.
[83] RINT, C. (ed.): *Handbuch für Hochfrequenz und Elektrotechnik*. 1st ed., Vol. 1 to 8. Hüthig & Pflaum, Heidelberg 1949 – 1969.
[84] RINT, C. (ed.): *Handbuch für Hochfrequenz und Elektrotechnik*. 12th, 13th ed., Vol. 1 to 5. Hüthig & Pflaum, Heidelberg 1978 – 1981.
[85] RINT, C. (ed.): *Lexikon für Hochfrequenztechnik, Nachrichtentechnik und Elektrotechnik*, Vol. 1 – 5. Porta, Munich 1957 – 1961.
[86] GEIGER, H., and SCHEEL, K., (ed.): *Handbuch der Physik*. Springer, Berlin 1929 – 1936.
[87] FLÜGGE, S. (ed.): *Handbuch der Physik*. Springer, Berlin-Heidelberg-New York 1956 – 1969.
[88] WIEN, W., and HARMS, F.: *Handbuch der Experimentalphysik*. Vol. 11, 2. Akadem. Verlagsgesellschaft, Leipzig 1935.
[89] WIEN, W., and HARMS, F.: *Handbuch der Experimentalphysik*. Vol. 16, 1. Akadem. Verlagsgesellschaft, Leipzig 1936.
[90] *Kleine Enzyklopädie Atom*. Verlag Chemie, Weinheim 1970.
[91] *Lexikon Elektronik*. Verlag Chemie, Weinheim 1978.
[92] *Handbuch der Elektronik*. Franzis, Munich 1979.
[93] *abc Optik*. Dausien, Hanau 1972.

Additional Literature

[94] ASHCROFT, N. W., and MERMIN, N. D.: *Solid State Physics*. Holt, Rinehart and Winston, Saunders College, Philadelphia 1976.
[95] BALLENTYNE, D. W. G., and LOVETT, D. R.: *A Dictionary of Named Effects and Laws in Chemistry, Physics and Mathematics*. Chapman & Hall, London 1984.

[96] *Berkeley Physics Course*. Vol. 1 – 6, McGraw-Hill, New York 1978 – 1984.

[97] BESANÇON, R. M. (Ed.): *The Encyclopedia of Physics*. Reinhold, New York, and Chapman & Hall, London.

[98] BORN, M., and WOLF, E.: *Principles of Optics*. Pergamon Press, Oxford-New York 1980.

[99] CALVEST, J. M., and McCLAUSLAND, M. A. H.: *Electronics*. Wiley, Chichester-New York 1978.

[100] *Encyclopedia of Physics*. Addison-Wesley, Reading, Mass., 1981.

[101] FEYNMAN, R. P., LEIGHTON, R. B., and SANDS, M.: *The Feynman Lectures on Physics*. Vol. 1 – 3, Addison-Wesley, Reading, Mass., 1963.

[102] FLOWERS, B. H., and MENDOZA, E.: *Properties of Matter*. Wiley, Chichester-New York 1979.

[102a] GLANSDORFF, P., and PRIGOGINE, I.: *Thermodynamic Theory of Structure, Stability and Fluctations.*. Wiley, London-New York-Sydney-Toronto 1978.

[103] GRANT, I. S., and PHILLIPS, W. R.: *Electromagnetism*. Wiley, Chichester-New York 1976.

[103a] HAKEN, H.: *Synergetics*. Springer, Berlin-Heidelberg-New York 1983.

[103b] HAKEN, H. (e.d.): *Springer Series in Synergetics*. 32 volumes. Springer, Berlin-Heidelberg-New York 1979 – 1986.

[104] HALL, H. E.: *Solid State Physics*. Wiley, Chichester-New York 1974.

[105] HALLIDAY, D., and RESNICK, R.: *Physics*. 3rd ed. Wiley, New York-London 1977.

[106] HECHT, E., and ZAJAC, A.: *Optics*. Addison-Wesley, Reading, Mass., 1979.

[107] HIX, C. F., and ALLEY, R. R.: *Physical Laws and Effects*. Wiley, Chichester-New York, and Chapman & Hall, London 1958.

[108] ITZYKSON, C., and ZUBER, C. B.: *Quantum Field Theory*. Mc Graw-Hill, New York 1980.

[109] JACKSON, J. D.: *Classical Electrodynamics*. 2nd ed. Wiley, Chichester-New York 1975.

[110] KITTEL, Ch.: *Introduction to Solid State Physics*. 6th ed. Wiley, New York-London 1986.

[111] KRANE, K.: *Modern Physics*. Wiley, Chichester-New York 1983.

[112] LONGAIR, M. S.: *Theoretical Concepts in Physics*. Cambridge University Press, Cambridge-London 1984.

[112a] MANDELBROT, B. B.: *The Fractal Geometry of Nature*. Freeman, New York 1983.

[113] MANDL, F.: *Statistical Physics*. Wiley, Chichester-New York 1978.

[114] *Mc Graw-Hill Encylopedia of Physics*. Mc Graw-Hill, New York 1983.

[115] MULLIGAN, J. F.: *Introductory College Physics*. Mc Graw-Hill, New York 1985.

[116] ORTON, J. W.: *Materials for the Gun Effect*. Mills & Boon Ltd., London 1971.

[116a] POSTON, T., and STEWART, I.: *Catastrophe Theory and its Applications*. Pitman, London-San Francisco-Melbourne 1978.

[117] ROWE, D. M., and BHANDARI, C. M.: *Modern Thermoelectrics*. Holt, Rinehart and Winston, London-New York 1983.

[118] SCHIFF, L. I.: *Quantum Mechanics*. Mc Graw-Hill, New York 1968.

[118a] SHU, F. H.: *The Physical Universe*. University Science Books, Mill Valley/California 1982.

[119] SMITH, F. G., and THOMSON, J. H.: *Optics*. Wiley, Chichester-New York 1975.

[120] WALD, R. M.: *General Relativity*. University of Chicago Press, Chicago-London 1984.

[121] WILLMOTT, J. C.: *Atomic Physics*. Wiley, Chichester-New York 1975.